P9-AON-577

GAMES AND SIMULATIONS IN
SCIENCE EDUCATION

Games and Simulations in Science Education

Henry Ellington, Eric Addinall
and Fred Percival

Kogan Page, London/Nichols Publishing
Company, New York

Copyright © Henry Ellington, Eric Addinall
and Fred Percival 1981

First published in Great Britain in 1981 and reprinted 1981 by
Kogan Page Limited, 120 Pentonville Road, London N1 9JN

British Library Cataloguing in Publication Data

Ellington, H I
    Games and simulations in science education
    1. Science — Study and teaching
    2. Teaching — Aids and devices
    I. Title   II. Addinall, E   III. Percival, F
    507'.8   Q185
    ISBN 0-85038-338-2

First published in the United States of America in 1981 by
Nichols Publishing Company, PO Box 96, New York, NY 10024

Library of Congress Cataloging in Publication Data

Ellington, Henry.
    Games & simulations in science education.

    Bibliography: p.
    Includes index.
    1. Science — Study and teaching. 2. Educational
games. 3. Simulation games in education.
I. Addinall, Eric, joint author. II. Percival, Fred,
joint author. III. Title.
Q181.E62 1980 507'.8 80-20748
ISBN 0-89397-093-X

Printed in Great Britain

Contents

Acknowledgements

The authors would like to acknowledge the assistance they received from the following in producing this book:

- Miss Jacquetta Megarry, of Jordanhill College of Education, for help at the planning stage;
- Mr Martyn Roebuck, HMI, of the Scottish Education Department, for checking the draft manuscript of Part 1;
- Miss Margaret Geddes, for typing the manuscript.

In addition, they would like to acknowledge the help they have received from the following during the seven-year programme of work on science-based games that made the book possible:

- Professor Norman Langton and Dr Peter Clarke, of RGIT, for continuous support and encouragement;
- Miss Margaret Geddes and Mrs Marjory Miara, for typing and secretarial help;
- Mr Michael Smythe, of the Institution of Electrical Engineers, for help in publishing and promoting their exercises.

Dr Percival would also like to acknowledge the help and support he received from his PhD supervisor, Dr Alex Johnstone, of Glasgow University, between 1973 and 1976.

The authors and publisher would like to thank the following for permission to reproduce copyright material:

- Chemical Teaching Aids (Letham, Ladybank, Fife, Scotland) for the Formulon sequences reproduced on pp 36 and 37;
- Ellis Horwood Ltd for Figures 5.1, 5.2, 5.5, 5.6 and 5.7, reproduced from *Interactive Computer Graphics in Science Teaching* (J McKenzie, L Elton and R Lewis, eds), Ellis Horwood Ltd, Chichester, England, 1978;
- Edward Arnold (Publishers) Ltd (41 Bedford Square, London WC1B 3DQ) for Figures 5.3 and 5.4, reproduced from the *HABER Unit on Ammonia Synthesis* (by R Edens and K Shaw) of the Chelsea Science Simulation Project;
- Heyden & Son Ltd (Spectrum House, Hillview Gardens, London NW4 2JQ) for Figures 3.1 and 3.2;
- Longman Group Ltd, Resources Unit (5 Bentinck Street, London W1M 5RN) for Figures 3.5, 3.6 and 3.7;

— Miss J Megarry, for the reproductions of Circuitron pieces on p 40 and for Figures 3.3 and 3.4.

Foreword

Because the authors work in the United Kingdom, this book is largely based on the British educational system. Everything in the book is, however, equally applicable to the educational systems of other English-speaking countries, particularly the United States, Canada, Australia and New Zealand. For the benefit of non-British readers, however, it might be useful to explain a number of 'local' terms which occur frequently and which may be unfamiliar.

First, the terms *primary, secondary* and *tertiary* as applied to the British educational system. In Britain, all children attend school from the age of 5 to 16, spending the first six or seven years in *primary schools* and the remainder in *secondary schools,* where they can, if they wish, stay on until they are 18 or so. All education after the secondary school stage is described as *tertiary education,* so that the tertiary sector includes all further education colleges, technical colleges, teacher training colleges, polytechnics and universities. Roughly speaking, British *primary schools* correspond to American *elementary schools, secondary schools* to American *high schools* and the British *tertiary sector* to the American *college, university* and *graduate school sector.*

Second, the terms A2, A4, etc as applied to paper sizes and the formats of leaflets, booklets and books. In Britain, the sizes of paper in most common use are submultiples of a basic size known as A1, which is roughly 84cm x 59.4cm. An A2 sheet is exactly half this size (59.4cm x 42cm), an A3 sheet one quarter of the size (42cm x 29.7cm), an A4 sheet one eighth (29.7cm x 21cm) and so on. The most commonly used sizes are A4 (the standard size used for writing paper and duplicating paper) and A5 (14.8cm x 21cm).

Introduction

During the last few years, a large number of science-based games, simulations and case studies have been developed, and these are now starting to be built into the curricula of our schools, colleges and universities. The use of such exercises seems certain to increase as more and more teachers, lecturers and curriculum designers become aware of their great potential. Until now, however, these developments have been hampered by the fact that there has been no basic text on science-based games, and no source book to which potential users could refer to find out what exercises were available in their particular field. This book has been written in an attempt to fill both these gaps.

The book consists of two main parts. Part 1 is a review of the potential role of games, simulations and case studies in science education and is made up of three sections. The first (Chapters 1 and 2) takes a broad look at the game/simulation/case study field, explaining what the various types of exercise are, showing how they are inter-related, discussing their general educational characteristics, and presenting a rationale for their use in science education. The second (Chapters 3, 4 and 5) examines the various types of exercise currently available to science teachers, dealing first with card and board games, then with other types of manual exercise, and finally with computer-based exercises; in each case, a general discussion of the main characteristics of the class is supported by detailed descriptions of specific examples. The third section (Chapters 6, 7 and 8) offers practical guidance on how to select and adapt exercises for particular purposes, how to use them in a teaching situation, how to design one's own exercises, and, finally, how to evaluate games, simulations and case studies.

Part 2 consists of a comprehensive collection of data sheets on science-based games, simulations and case studies — the first such collection to be published. It includes all exercises known to the authors that are (a) suitable for use in science education, (b) generally available to teachers, and (c) written in English.

The data sheets are arranged alphabetically in four separate sections, which contain respectively physics-based exercises, chemistry-based exercises, biology-based exercises and other exercises with some science or technology content. Readers who know of any science-based games, simulations or case studies which are *not* included are asked to let the authors know (preferably by sending completed data sheets), so that such exercises can be included in future editions of the book.

With regard to the way in which references have been incorporated into the text of Part 1, the aim has been to produce a free-flowing introductory book for practising science teachers rather than a rigorous academic treatise. For this reason, references are given as an unobtrusive series of superscript numbers rather than by interrupting the script with names of authors and dates of papers, books, etc. Readers who wish to study any topic in greater depth can follow up the appropriate references, which are listed in full at the end of the book, or consult the bibliography that follows the reference section.

Part 1:
The potential role of games and simulations in science education

Chapter 1

General introduction to games and simulations

Since the use of gaming and simulation techniques in science education is a recent development, it is likely that many readers will be unfamiliar with the field. This opening chapter will therefore set the scene for what is to follow by explaining what games and simulations are, describing their main educational characteristics, and reviewing the history of their use in education and training to date.

The game/simulation/case study field

Until the early 1970s, there was a great deal of confusion over what exactly was meant by the terms 'game' and 'simulation' when used in an educational context. Many authors, for example, regarded the terms as being largely synonymous, while others regarded games as being a particular type of simulation. This confused situation was greatly clarified in 1973, when Bloomer[1] presented the thesis that games and simulations are in fact quite different types of exercise, and suggested what was to form the basis of a workable terminology and classification.

Bloomer recognized that games and simulations have distinct, but not incompatible, properties, and that the set of games overlaps the set of simulations to form a hybrid set of exercises which she called simulation games (see Figure 1.1).

The key to Bloomer's classification was her choice of definitions for games and simulations. For the former, she adopted the definition

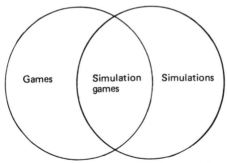

Figure 1.1: The relationship between games and simulations (after Bloomer)

previously given by Abt,[2] which states that a *game* is 'any contest (play) among adversaries (players) operating under constraints (rules) for an objective (winning, victory or pay-off)'. This definition identifies two essential features which must *both* exist before an exercise can reasonably be described as a game. First, it must involve *overt competition* of some sort, either directly between players (as, for example, in bridge) or between individual players who are competing against the 'game system' (as, for example, in golf). Second, the exercise must have *rules*, ie the players must operate under a set of arbitrary constraints specific to the particular game. One possible weakness of the definition is that it appears to exclude from the class of games exercises in which a single player competes directly against the game system (patience, solitaire, crosswords, fruit machines, etc); this anomaly can be resolved, however, by regarding the deviser of the game system as one of the 'adversaries' in such cases. (Somewhat more disturbingly, it could be argued that the definition includes most modern wars, since these undoubtedly involve competition and are — in theory at any rate — fought under the rules of the Geneva Convention; the authors challenge readers to convince them that such wars do *not* fall into the class of activities circumscribed by Abt's definition!)

In the case of simulations, Bloomer adopted Guetzkow's[3] definition which states that a *simulation* is 'an operating representation of central features of reality'. This definition again identifies two essential features which must *both* exist before an exercise can reasonably be described as a simulation. First, it must represent a *real situation* of some sort. Second, it must be *operational*, ie it must constitute an *ongoing process*. As shown by Bloomer, this latter criterion effectively excludes from the class of simulations static analogues such as photographs, maps, graphs and circuit diagrams, but includes working models of all types.

It immediately follows from the above definitions that a *simulation game* is an exercise that possesses the essential characteristics of both games (competitions and rules) and simulations (ongoing representation of real-life), examples being chess and Monopoly. The term is now generally accepted by workers in the gaming/simulation field.

An even more recent development has been the recognition that case studies can be related to the field of games and simulations. Such exercises have been used for many years as a teaching method in their own right. Their use in training students for the medical and legal professions is well known, and they have been an established teaching technique in business and management studies for many years.

In 1974, Walker[4] defined a *case study* as 'a study in detail of a particular event, problem or situation'. Apart from the fact that it is partly circular, the authors believe that this definition fails to identify one extremely important general characteristic of case studies, namely that they are invariably carried out either in order to identify special features of the case under study or in order to illustrate general characteristics of the broader set of situations of which it is typical.[5] (Under Walker's definition, reading a chapter of any book that gives a detailed description of its subject matter would constitute carrying out a 'case study'.) The authors prefer the following extension of Walker's definition, in which a case study is defined as 'an in-depth examination of a real-life or simulated situation carried out in order to illustrate special and/or general characteristics'.[5]

The relationship that exists between case studies and the simulation/ gaming field was recognized by Reid,[6] who, in 1977, argued that interactive case studies have many features in common with simulations, and that the set of case studies therefore overlaps that of simulations in the way shown in Figure 1.2.

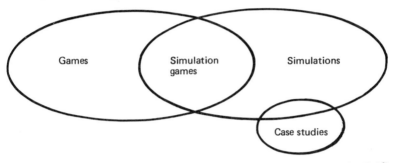

Figure 1.2: The relationship between games, simulations and case studies (after Reid)

Although Reid's picture represents possibly the most common area of overlap between games, simulations and case studies, the authors would go even further and would contend that the set of case studies also overlaps the sets of simulation games and games, as shown in Figure 1.3.[5]

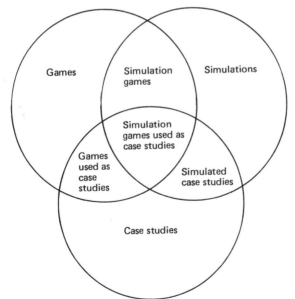

Figure 1.3: The relationship between games, simulations and case studies according to the authors

If this view is accepted, it follows that there are no less than seven distinct types of exercise in the game/simulation/case study field, namely, three 'pure' types (games, simulations and case studies) and the four 'hybrid' types that occupy the various areas of overlap in Figure 1.3. In order to help readers appreciate the distinctions between the various types, we will now give specific examples of exercises that fall into the different areas of the figure.

'PURE' GAMES

This class contains all exercises which have the two basic characteristics of games (*competition* and *rules*) but lack some or all of the basic features that characterize simulations and case studies respectively. It includes well-known games such as Scrabble and Othello as well as the majority of conventional card and ball games.

'PURE' SIMULATIONS

This class contains all exercises which have the essential properties of simulations (ie represent a *real situation* and are *ongoing*) but lack some or all of the essential characteristics of games and case studies. It includes all the various mechanical/electronic simulators that are used to train future operators of complex machinery, aircraft, weapon systems, and so on, one of the best-known examples being the Link Trainer developed during the Second World War to teach basic flying skills.

'PURE' CASE STUDIES

This class contains all exercises which have the essential features of case studies (*in-depth study* and illustration of *special* or *general characteristics*) but lack some or all of the basic features that characterize games and simulations. It includes conventional non-interactive case studies of the type carried out in the training of doctors and lawyers (the type of exercise after which the class was named) as well as similar exercises carried out in other disciplines. The series of Jackdaw packs (collections of facsimiles of documents, drawings, contemporary records, etc used in the study of specific historical periods of events) belong to the class, as do more complicated exercises such as What Happens When The Gas Runs Out? and The Sulphuric Acid Story (see Chapter 4 and Part 2).

SIMULATION GAMES

This class contains those exercises which have the basic characteristics of both games and simulations but not of case studies. It includes many popular family games, such as Monopoly and Cluedo, as well as the great majority of commercially available war games (Campaign, Bonnie Prince Charlie, etc). Possibly the most familiar example of the class, however, is chess, which is believed to have originated in sixth-century Persia or India as a simulation of a contemporary battle (how many people realize that the movements of the rook represent those of the war elephant?).

GAMES USED AS CASE STUDIES

This class contains exercises which have the basic properties of games and case studies but not of simulations. Probably the best-known examples are the various games developed at the University of Michigan by Layman Allen as aids to the teaching of symbolic logic and mathematics — Wff'n Proof, On-Sets, Equations, etc. (Dealing as they do with purely formal fields, such exercises cannot be said to represent 'reality' and must, therefore, be excluded from the set of simulations.) Other examples are the use of Mastermind as a source of case studies in logic and problem solving,[7] and the use of gambling games, such as coin tossing, in the study of probability theory. Many of the chemistry-based card games described in Chapter 3 and Part 2 also belong to the class.

SIMULATED CASE STUDIES

This class contains exercises which have all the essential properties of simulations and case studies but not of games. It includes, for example, the 'simulated patient' technique developed at McMaster University, Canada for use in the training of doctors in diagnosis and treatment. Here, trained 'actors' take the place of real patients, thus allowing trainee doctors to learn from their mistakes without any risk of the serious (or even fatal) consequences which could result from such mistakes in real life. The class also contains a wide range of computer simulations in fields such as physics, chemistry, engineering and economics, the extended study of the world's possible future development carried out by the Club of Rome being a particularly well-known example.[8] It also includes many of the exercises described later on in this book, eg Power for Elaskay (see Chapter 4 and Part 2).

SIMULATION GAMES USED AS CASE STUDIES

This class contains those exercises which have all the essential characteristics of games, simulations and case studies, and covers an extremely wide range of types. For example, it includes many of the small-scale simulation games which have been designed for use in specific teaching situations, eg most of the science-based exercises published by Longman (see Chapter 3 and Part 2). It also contains large-scale interactive teaching exercises such as The Power Station Game, The Amsyn Problem and Fluoridation? (see Chapter 4 and Part 2). All the 'real' war games carried out in military training also fall into this class (see below), as do many computerized business management games such as the Bruce Oil Management Game.[9]

Why games and simulations are educationally useful

Now that we have seen what games and simulations are and shown how they are related to case studies, let us examine some of the reasons why such exercises are claimed to be useful from the educational point of view.

1. Games and simulations constitute a highly versatile and flexible medium whereby a wide range of educational aims and objectives can be achieved.[10] They can be used to achieve objectives from all parts of Bloom's cognitive[11] and affective[12] domains, and simulations can also be used very effectively in the psychomotor area. It has been found that gaming and simulation techniques are no more effective than other, more traditional, methods when used to teach the basic facts of a subject,[13] but are particularly useful for achieving high-level cognitive objectives relating to such things as analysis, synthesis and evaluation and also for achieving affective objectives of all types.[10] Thus, it is not advocated that they be employed as a main, front-line teaching technique, but rather as a complement and support to traditional methods, eg for reinforcement or to demonstrate applications or relevance.

2. The use of a simulated as opposed to a real situation as the basis of an educational exercise allows the situation to be tailored to meet the needs of the exercise rather than requiring the exercise to be designed within the constraints imposed by the situation. Only very rarely does a real-life situation have all the features that the designer of an educational exercise wishes to highlight or bring out, whereas a simulated situation can have all such features built in. Also, real-life situations are often far too complicated to allow them to be used as the basis of an educational exercise as they stand; the simplification that the use of simulation allows can often overcome this difficulty by reducing the complexity to manageable proportions.

3. Research indicates[14] that well-designed games and simulations can achieve positive transfer of learning — the ability of participants to apply skills acquired during the exercise in other situations.[15] It would, in fact, be difficult to justify the use of many simulation-type exercises if no such transfer of learning occurred. The whole point of the various types of cockpit simulator, for example, is to enable aircrew to acquire the various skills needed to operate a particular type of aircraft in a safe, controlled environment and then to be able to use these skills when flying the actual aircraft. The same is true of virtually all other training simulators.

4. In many cases, games and simulations constitute a vehicle whereby participants can use and develop their initiative and powers of creative thought. This feature of games and simulations could prove increasingly important if our educational system continues to place progressively greater emphasis on the cultivation of divergent thought processes.

5. Apart from their purely cognitive, content-related outcomes, many games and simulations help foster a wide range of non-cognitive skills (such as decision making, communication and inter-personal skills) and desirable attitudinal traits (such as willingness to listen to other people's points of view or appreciate that most problems can be viewed in a

number of different ways).[6, 16] Indeed, some workers[6] believe that this is the area in which games and simulations are capable of making their most valuable contribution to education. Exercises that involve interaction between the participants are especially effective in this regard.

6. One advantage which games and simulations appear to have over more traditional teaching methods is that pupil or student involvement is normally very high — a feature that is particularly beneficial for the less able. In addition, most participants find games and simulations extremely enjoyable.

7. In cases where a competitive element is included (not necessarily at the expense of co-operation), many workers believe that this provides strong motivation for the participants to commit themselves wholeheartedly to the work of the exercise.[17, 18] Such a competitive element may be overt (when groups or individuals are in direct competition with one another) or it may be latent (as, for example, in exercises like Power for Elaskay where groups or individuals have to perform parallel activities and then report on their findings); in both cases, the authors have no doubt that motivation is increased.

8. Many games and simulations have a basis in more than one academic discipline, a feature that can help the participants to integrate concepts from otherwise widely related areas into a cohesive and balanced 'world picture' — surely one of the main overall aims of any worthwhile educational system. Exercises which require the participants to formulate value judgements or examine technological problems from other than a purely scientific point of view are claimed to be particularly valuable in this respect.[19]

9. Another valuable characteristic of multi-disciplinary exercises is that they often provide a situation in which participants with expertise in different subject areas have to work together efficiently and harmoniously in order to achieve a common end. Inter-personal skills of this type are vitally important for success in later life and constitute an area of education in which the multi-disciplinary simulation and simulation game are virtually the only means of providing practical experience in a classroom or college environment.[10]

Review of use of games and simulations to date

Games have been played for amusement for thousands of years, and simulation (in its broadest sense) has an equally long history. The application of simulation and gaming techniques to education and training is, however, a comparatively recent development. The first field in which such applications took place was military training. Here, serious use of simulation and gaming began at the end of the eighteenth century, and the techniques have since been developed to a high level of realism and sophistication. The next field in which important developments took place was business management training, where the use of gaming and simulation as a means of developing decision-making skills was introduced in the mid-1950s. It was not until the early 1960s, however, that the use of such techniques spread to secondary and tertiary education, with the initial developments taking place in teacher training and in the social sciences. Let us now examine these various developments in more detail.

MILITARY APPLICATIONS

It is sometimes argued that chess (developed in the sixth century) was the first war game, but it was not until 1798 that a game involving the use of maps was used in military training.[20] Over the next century, two basic types of war game came into increasingly regular use.[21] The first of these, the so-called 'rigid' war games, simulated realistic situations, the changing pattern and fortunes of war being introduced by throwing dice. These formal academic exercises could be highly abstract, their main use being in the development of skills relating to overall strategy. The second type, the so-called 'free' games, involved the use of actual military units in field situations, and were used to develop tactical skills. The two types of game were therefore complementary. Although their early use was largely confined to Prussia (where they were first developed), both types of game have since been adopted by practically every major military force in the world.

BUSINESS MANAGEMENT APPLICATIONS

Games and simulations had been in regular use by the military for over 150 years before their next major application was found, namely in management training. Here, there was a need to find a teaching method which could bridge the gap between formal, academic instruction (which often lacked direct job relevance) and on-the-job training (which could be slow, and was generally restricted to a limited area). Around 1955, it was recognized that gaming and simulation methods could help provide a solution and, in 1956, the American Management Association produced the first business game.[22] This was a decision-making simulation exercise for potential executives.

The idea of using gaming and simulation techniques in business and management training quickly spread throughout the western world and, in response to the rapidly increasing demand, a wide range of materials soon became available. These ranged from relatively simple manual exercises (exercises based on the use of conventional printed materials such as booklets and work sheets) and board games to highly sophisticated computer-based simulations, some being co-operative in nature, and others highly competitive. The use of such techniques is now an integral part of business management training at all levels from school to in-service industrial development.

EDUCATIONAL APPLICATIONS

One of the first areas outside the military and business fields in which gaming and simulation methods were used was teacher training. Here, the first simulations were published in 1962;[23, 24] these required future teachers to deal with a range of 'everyday' situations in a simulated school. Other exercises of a similar type soon followed.[25]

Although most of the early educational developments in the field of gaming and simulation were of American origin, materials started to be produced in Europe in the late 1960s — mainly for use in the teaching of social sciences such as geography,[26] international relations[27] and urban development.[28] During the 1970s, however, gaming and simulation techniques spread to an ever-increasing range of other subjects, and from about 1970 onwards, they started to be used in the teaching of science. This last development will be the subject of the next chapter (and, indeed, the remainder of the book).

Chapter 2

How games and simulations can be used in science education

In the first part of this chapter, the various ways in which the authors believe that gaming and simulation techniques are capable of making a significant contribution to science education will be discussed in detail. The second part will give a historical review of some of the more important developments that have taken place to date.

Rationale for the use of science-based games and simulations

There are three ways in which science-based games and simulations can be used in secondary and tertiary education, namely:

(a) as aids to the teaching of the basic content of science courses;
(b) for educating *through* science (using science-based exercises to cultivate useful skills and desirable attitudinal traits);
(c) for teaching *about* science and technology and their importance to modern society.

Note that these are not mutually exclusive, since many exercises are capable of contributing to more than one area. Let us now examine each in more detail.

(a) *How games and simulations can be used in the teaching of science.* In Chapter 1, it was claimed that games and simulations are capable of achieving a wide range of educational aims and objectives but are no more effective than traditional methods in teaching the basic facts and principles of a subject. Because of this, it is advocated that they be used not as a front-line teaching method, but rather in a complementary and supportive role. There are two main ways in which this can be done.

(i) *For reinforcing basic facts and principles.** Once the basic facts of a particular section of a course have been taught, it is often necessary to reinforce the knowledge the pupils or students have just acquired by giving them some form of exercise in which they have to demonstrate their understanding of what they have learned by applying it to a specific situation. Traditionally, such exercises have generally taken the form of worked examples the pupils or students complete on their own, either in class or at home. In many cases, however, it would be possible to achieve the same objectives by

---

* Throughout this book, the terms 'reinforce' and 'reinforcement' are not used in their strict psychological sense; rather, they refer to an improvement in the ability to recall or apply facts or principles.

making use of a game or simulation of some sort and, as we saw in Chapter 1, this would probably have the added advantage of increasing motivation (if only by introducing some variety into the course). It is certainly not recommended that *all* worked examples in a course should be replaced by games or simulations, nor even that a high proportion should be. Rather, it is suggested that it might, in some cases, be beneficial to introduce a few carefully chosen games or simulations into a course *in situations where they seem to offer some distinct educational advantage over more traditional alternatives.*

The types of exercise that are probably best suited to this form of application are fairly short games, simulation games and simulated case studies with strictly limited objectives. The various chemistry card games (Chemsyn, Formulon, etc) are good examples, as are board games such as Circuitron and the exercises published by Longman (see Chapter 3 and Part 2). Another good example is The Young Chemist, a game developed at the Israel Institute of Technology, Haifa, in order to help senior schoolchildren master the complexities of the periodic table.[29]

(ii) *For developing laboratory skills.* Possibly the most important potential application of games and simulations in the teaching of science is as a supplement to and, in some cases, a substitute for conventional experimental work. One of the first exercises to be used in this way was Circuitron, which was specially developed to help pupils apply basic circuit theory in Scottish schools.[30, 31]

It is, however, simulations rather than games — and, in particular, computer-based simulations — that offer the most exciting possibilities in this area.[32, 33, 34] Here, the imminent prospect of cheap, highly versatile microcomputers becoming generally available in schools and colleges means that it will be possible to give pupils and students direct experience (through simulations) of a far wider range of experimental situations than has been feasible up to now. Specific areas in which such simulations could make a significant contribution include the following:

— situations where a conventional experiment is either extremely difficult or impossible (eg experiments in astrophysics and human genetics, and 'thought experiments' such as the investigation of non-inverse-square or negative gravitation);

— situations where experimental apparatus is either not readily available or too complicated or expensive for general laboratory use (eg experiments in high-energy physics, nuclear physics or reactor physics and industrial processes of all types);

— situations where actual experimental work could be dangerous (eg work with explosive mixtures, highly radioactive materials, highly toxic chemicals, virulent pathogens, etc);

— situations where a conventional experiment would take an unacceptably long time to complete (eg experiments in genetics and population dynamics or with long-lived radioactive materials).

All these potential applications of computer simulations will be discussed in much greater detail in Chapter 5.

(b) *How games and simulations can be used to educate 'through' science.* A major theme of current thinking regarding the development of science education at both secondary and tertiary levels is the identification and achievement of the various 'desirable' skills, habits, attitudes and modes of thinking which the end products of that education should possess over and above purely factual knowledge. Since the 1930s, there has been an increasing awareness of the importance of including in science courses objectives which are not purely cognitive but, until comparatively recently, there was little evidence of any serious attempt being made to do so.[35] Since the mid-1960s, however, a number of new science schemes developed for Scottish and English secondary schools have incorporated such broader objectives, and a similar trend has recently been noticeable at tertiary level.[36, 37, 38]

This change of emphasis in science education has not only caused teachers and lecturers to re-think their approach regarding how broader objectives can best be achieved, but has also caused many to go further, and ask the fundamental question: 'What is a science education for?'[6]

The purely vocational advantages of teaching science are self-evident in the case of senior undergraduates bent upon a scientific career or technicians being trained in technical colleges or polytechnics. In the case of the bulk of those to whom science is taught (secondary schoolchildren and junior undergraduates), such advantages are less obvious, however, and the various other arguments put forward for teaching science are, in general, unconvincing and lacking in empirical support.

One such argument is that all future citizens should have at least a basic knowledge of science in order to help them cope effectively with the highly technological world in which they live. This argument, although at first sight convincing, has at least two basic weaknesses. First, there is no real evidence that the type of cognitively orientated science traditionally taught in schools and colleges is of any significant use in helping students make sense of their environment. (It is for this reason that 'science in society' courses of the type discussed below are now being developed.) Second, there is nothing to suggest that non-scientists are any less successful than scientists in coping with life.

Another argument is that a scientific training helps students think logically and dispassionately about everyday problems and approach them by means of the 'scientific method'. Research carried out in secondary schools has, however, indicated that this argument may have little foundation in the case of our present science courses.[39] Other research has shown that, in most cases, science courses appear to have no substantial influence on the development of pupils' attitudes to science, and that few teachers make any serious effort to achieve goals specifically related to attitude and interest.[40] Thus, it appears that the hoped for 'spin-off' effects supposedly associated with science courses are not necessarily achieved when traditional teaching methods are employed.

These results are not particularly surprising when one considers that the great majority of traditional science courses at all levels concentrate almost exclusively on the inculcation of cognitive and psychomotor skills, with the vague hope that the various non-cognitive skills (eg decision making, communication, problem solving, library and inter-personal skills) and desirable attitudinal traits, supposed to be a 'bonus' from a science education, will somehow 'rub off' on the students.[10] All too often, little or no attempt is made to cultivate such non-cognitive outcomes of science courses. There is, however, a strong case for making a conscious effort to foster their development, since even a professional scientist is unlikely to succeed in our complex and changing society purely on the basis of cognitive attainment. If we accept the description of education as 'what is left when the facts have been forgotten', the case becomes even stronger.

Since traditional teaching methods have proved to be of doubtful efficiency in achieving non-cognitive outcomes of the type described above, it is obvious we will have to develop new methods that are specifically designed to achieve such outcomes (assuming, of course, that we accept the argument that they are educationally desirable). It is the contention of the authors[16, 41] that participative science-based simulation games and simulated case studies are capable of fulfilling this role. If properly designed, these provide a means of educating 'through' science, ie of using a science-based exercise as a vehicle for achieving a wide range of educational objectives going far beyond those that would normally be associated with its intrinsic scientific content.

Many of the exercises described in this book (including nearly all those developed by the authors and by Norman Reid) have been designed with this type of usage in mind. For example, Proteins as Human Food was developed to foster communication and inter-personal skills in senior schoolchildren and junior undergraduates,[42] while Polywater was developed to help senior chemistry undergraduates acquire library skills.[43] Further details about exercises that can be used for education 'through' science are given in Chapter 4 and Part 2.

(c) *How games and simulations can be used to teach* about *science and technology.* Since 1959, when C P Snow gave the now famous Reith lecture in which he presented the doctrine of the 'two cultures',[44] there has been a growing realization of a number of fundamental deficiencies in our educational system — particularly in relation to science education.

First, at an academic level, there is the undoubted dichotomy between scientists and technologists on the one hand and non-scientists (particularly those with a background in humanities such as literature) on the other.[44] This is a direct result of the early specialization that takes place in our educational system, which means that the majority of our schoolchildren study no science at all after the age of 15 or so, while the minority who opt for science courses study precious little else for the rest of their academic careers. As C P Snow has pointed out, this has led to a virtually complete lack of understanding between the two groups and, in many cases, has given rise to mutual mistrust and even hostility.

Second, at a more general level, there has been an almost complete failure to make our future citizens aware of the relevance of science to the real world and of the vital importance of science and technology to modern society.[45, 46] Throughout their adult lives, these citizens are required to make political, social and economic decisions, many of which have a scientific basis or component and, at the moment, our educational system is not preparing them to make such decisions in a reasoned and informed manner. In other words, our society lacks what Bernard Dixon, editor of *New Scientist,* describes as 'communal technical literacy'.[47]

The above problems are likely to become even more acute during the 1980s as microprocessor-based technology comes to play an increasingly important role in all our lives.[48] It has been predicted that the microprocessor will have an impact on society comparable to that produced by the development of the steam engine during the latter part of the eighteenth century, but that this second industrial revolution will take effect much more rapidly than the first. The microprocessor seems likely to bring about fundamental changes in nearly every facet of modern society and, if these changes are to take place without causing complete social upheaval, many commentators believe that it will be *essential* to develop educational programmes capable of producing technically literate citizens as soon as possible.[48, 49]

A promising start has been made by the recent development of 'science in society'-type courses at both secondary and tertiary level,[50] and it is now generally recognized that games and simulations are capable of making a significant contribution to courses of this type.[45, 51, 52] The authors believe that science-based simulation games and simulated case studies, particularly those of a multi-disciplinary nature, are ideally suited to such a role, since they represent one of the most effective means at our disposal of demonstrating the role of science and technology in modern society.

One great advantage of exercises of this type is that they can be incorporated equally successfully into both science and non-science courses. With the former, they can either be built into the main fabric of the course as supportive case studies, or they can be made part of the 'mind-broadening' associated studies element now becoming an integral part of most of our secondary and tertiary science courses. With the latter, they can again be made part of the general studies element of the course — possibly by watering down or removing the 'hard science' content in order to convert them to a form that can be handled by people who lack a technical or mathematical background. The authors have found that many exercises originally designed for use in science courses can, by suitable modification, be used effectively in this way. A good example is The Power Station Game (see below), which was originally developed as an exercise for physics students of roughly A-level standard,[53] but which is now being used with a wide range of other groups, including trainee primary teachers.[54]

Detailed information about other exercises that can be used to demonstrate the social relevance of science and technology can be found in Chapter 4 and Part 2.

Review of early developments

Now that we have established how science-based games and simulations can contribute to our educational system, let us examine some of the developments that have taken place in these areas since the first exercises began to appear in 1970.

(a) *Developments in the use of games and simulations for teaching science.* The first games and simulations to be specifically developed for use in the teaching of science were a series of card and board games published between 1970 and 1975.

Among the earliest of these were a number of relatively simple games intended for use at secondary school level. In Britain, they included Chemsyn and Element Cards (published by Heyden and Son Ltd), Formulon and Properties and Substances (published by Chemical Teaching Aids) and Ionics (published by Science Systems Ltd).[55] In America, they included Elements, Compound, The Geologic Time Chart Game and Lab Apparatus (all published by Union Printing Co Inc) and a range of chemistry-based games published by Teaching Aids Co. All these games were designed for use in the reinforcement of basic facts, concepts and principles, and all were short enough to be fitted into the normal teaching curriculum without causing undue disruption. (See Chapter 3 and Part 2.)

At roughly the same time, a somewhat more ambitious exercise — Circuitron — was developed in the Education Department at the University of Glasgow for inclusion in Project PHI,[56] a package of programmed materials for use in teaching science in small, remote secondary schools in the Highlands and Islands of Scotland. Circuitron (which will be examined in detail in Chapter 3) was a board game designed for use as a supplement to conventional experimental work on electrical circuits, and could be used at a variety of levels ranging from lower secondary to tertiary.[31, 33] It was published by Griffin & George in 1972 and was one of the first exercises of its type to receive a thorough evaluation.[31]

Three years later, a package of 18 games, covering a variety of topics in physics, chemistry, biology and general science, was published in Britain by Longman. The games were all short exercises intended for use at secondary level for reinforcing basic concepts and principles, and were again designed to fit easily into the school curriculum. Like the parallel series of geography-based games produced by the same company, they were published in the form of booklets containing all the resource materials (board, cards, tokens, etc) needed for a single playing group (between 4 and 12 pupils, depending on the game). Typical examples were The Great Blood Race, dealing with the composition, circulation and physiological functions of human blood and Competition Among The Metals, which helped to show how different metals have different reactivities. (See Chapter 3 and Part 2.)

By this time, the first computer-based simulations were also starting to be developed.[33] Establishments in which early work took place in Britain included the IBM Research Centre at Peterlee,[32] the Centre for Science Education at Chelsea College[57] and the Institute for Educational Technology at the University of Surrey.[58] More recently, nearly every educational establishment with access to a computer has developed its own range of computer exercises, but the vast majority of these are not generally available. Computer simulations will be dealt with in greater detail in Chapter 5.

(b) *Developments in the use of games and simulations for education 'through' science.* The first developments in this area again took place during the early 1970s, and it is interesting to note that most of these early exercises were concerned with ecological or environmental topics.

At the Universities of Sussex and Bath, for example, a number of ecology-based games and simulations were built into degree courses in biology.[59, 60] Subsequently, four of the Bath exercises were adapted for use at secondary-school level and published as a multi-media package under the title of The Ridpest File (see p 190). This was one of a number of similar packages developed by the School of Education at Bath for organizations such as BP; these all dealt with the interface between industry and the environment.

At the same time as this work was being carried out in England, a number of important developments were taking place in Scotland and America. The first was the Star River Project, a large-scale role-playing exercise that was developed by the Clyde River Purification Board in collaboration with CRAC (the Careers Research and Advisory Centre) for use at post-graduate level. This was based on a multi-disciplinary scenario involving biology, chemistry, geography and economics. In 1972 the exercise was modified for use at sixth-form level in schools, and the final package was published by Esso Petroleum Co in 1973 (now discontinued). A similar exercise (The Dead River, see p 189) was developed in America at roughly the same time and was also published in 1973 (by Union Printing Co).

As a follow-up to the Star River Project, the Scottish Education Department set up a working party in 1973 in order to develop a second large-scale multi-disciplinary exercise for use at sixth-form level, this time with a basis in physics. The work was carried out in Robert Gordon's Institute of Technology, Aberdeen, where a team of Institute staff, local physics teachers and educational administrators developed The Power Station Game, a role-playing exercise based on the design of a large power station.[53] Following extensive field trials in and around Aberdeen, the final version of the game was published by the Institution of Electrical Engineers (IEE) in 1976. It has since been used in a wide variety of educational and training situations,[53, 54, 61] and has also generated a number of further exercises, most of which have had a fairly high physics or engineering content (see Chapter 4 and Part 2).[53, 62]

While The Power Station Game and its successors were being developed in Aberdeen, parallel work in chemistry and biology was taking place in Glasgow. This was co-ordinated by the Science Education Group of the

University of Glasgow, where a whole range of 'mind-broadening' simulation games and simulated case studies, designed for incorporation into chemistry courses at both secondary and tertiary levels, were developed between 1973 and 1978.[43, 63, 64] All were thoroughly field tested and evaluated,[10, 65] and all have subsequently been published — three by the Education Division of the Chemical Society and the remainder by the Scottish Council for Educational Technology. Some of these exercises will be examined in detail in Chapter 4, and all are described in Part 2.

A similar series of biology-based games, designed for use at secondary-school level, was also developed at Glasgow. Although these were all field tested and evaluated,[66] they have not so far been published.

One of the most interesting recent applications of science-based simulations has been their use as the basis of a number of inter-school competitions. The first of these, Hydropower 77, was organized by the North of Scotland Hydro-Electric Board (NSHEB) during the winter of 1976-77 for secondary schools in their area.[67] It was based on a highly demanding simulated case study in which cross-disciplinary teams of senior pupils had first to design a hydro-electric pumped storage station and then write a 'consultant's report' and prepare a multi-media presentation on their proposed scheme. The competition project was subsequently converted into a self-contained teaching package and published by the IEE under the name Hydropower (see Chapter 4 and p 13

During the following winter, the NSHEB ran a further competition of a similar type, this time based on alternative energy.[69] The competition project was also subsequently converted into a self-contained teaching package, which was published by the IEE under the name Power for Elaskay (see Chapter 4 and p 139).[70, 71]

A third inter-school competition, known as Project Scotia, was run by the IEE during the winter of 1978-79 — this time on a UK-wide basis.[72] The competition project, which involved designing a UHF television broadcasting network for a remote area, was developed in conjunction with the BBC and the Independent Broadcasting Authority (IBA). It will eventually be published as a self-contained teaching package for use at secondary and tertiary levels.

The IEE has also organized a series of regional competitions based on The Power Station Game. The first was held in Aberdeen in 1977,[73] the second in Brighton in 1978[74] and the third in Lancaster in 1979.

(c) *Developments in the use of games and simulations for teaching about science and technology.* Although the main purpose of the various games, simulations and competitions described in the previous section was to use science-based exercises as vehicles for achieving non-cognitive objectives (eg cultivating communication, decision making and inter-personal skills), they were, in almost every case, also designed to teach the participants about the social relevance of science and technology. Thus, virtually all the exercises could be said to have been designed for teaching *about* science and technology as well as for educating *through* science.

Many have since been used for this purpose, both with science and engineering students and with non-science students.[54, 61, 75]

Perhaps the most important development in this area has been the Association for Science Education's Science in Society course. This is an alternative O-level (AO level) course in science, designed to be taken in the sixth form, as a complement to conventional science and arts courses.[45, 51] Work on the project started in 1977, and it was decided at an early stage to make extensive use of games and simulations in the course.[76] Furthermore, these were built into the basic fabric of the course curriculum rather than used as 'optional extras', as had so often been the case with such exercises in the past. A series of simulation games and simulated case studies was specially developed by the project team and, following exhaustive field trials and modifications, was published as part of the overall course package (see Part 2).

Chapter 3

Card and board games

In this and the following two chapters, the various types of game, simulation and case study that can be used in science education will be examined in depth. This chapter will deal with card and board games, Chapter 4 with other manual exercises, and Chapter 5 with computer-based exercises. In each case, a general discussion of the main educational characteristics of the different types of exercise will be illustrated by detailed examination of specific examples.

Card games

The class of 'card games' is taken to include all games where the playing materials consist solely of cards of one form or another. Games (such as Circuitron) which also make use of a board are excluded, such exercises being regarded as belonging to the class of 'board games'. As we saw in Chapter 2, the class contains some of the first science-based games to be developed, including well-known examples such as Chemsyn, Formulon and Ionics.

Card games have the following general characteristics:

(a) They are invariably compact and are thus easily stored and carried around, as well as being relatively inexpensive.

(b) Most card games are simple, with comparatively short playing times. This enables them to be fitted easily into the structure of most curricula, either by building them into the actual lessons as reinforcing exercises or by inviting pupils or students to play them during lunch breaks, free periods, etc.

(c) The great majority of educational card games are based on the formats of well-known games such as rummy, solitaire or dominoes. This means that most pupils and students find their rules easy to pick up.

(d) Most educational card games are fun to play and, in exercises of an interactive nature, the competition factor is generally high. This increases motivation in two ways, since each player
  (i) wants to ensure that what he does is correct, so that he can play to maximum effect;
  (ii) wants to find fault in what other players are doing, thus ensuring that their efforts are subjected to continuous critical scrutiny.

(e) Card games are generally limited to relatively low-level cognitive objectives (eg reinforcement of knowledge, fostering of understanding, demonstration of simple applications), although they can also be useful in cultivating simple decision-making skills. Provided their intrinsic limitations are fully appreciated by both designers and users, however, this need not be a disadvantage.

(f) A somewhat more serious potential weakness of card games is the danger that, unless the game structure and educational content are properly integrated, pupils and students can play them *purely as games* without deriving any real educational advantage from so doing. (A number of commercially available science-based games tend to have this weakness to a greater or lesser extent.[55]) In designing an educational card game, it is essential to ensure that players have to use the *educational content* in order to be able to play effectively, and not simply the *structure* (ie the rules). Provided this is done, a card game can constitute a highly effective learning situation.

Let us now examine two specific examples of science-based card games, namely Formulon and Chemsyn.

FORMULON

Formulon (p 159) is designed for use in the teaching of basic inorganic chemistry at lower and middle secondary-school level (ie with pupils aged roughly 13 to 16). The game consists of a pack of 100 cards plus a three-page instructional leaflet; 78 of the cards represent atoms or ions, 20 multipliers (2 or 3) and the remaining 2 Mendeleef cards (jokers) can be used to represent any atom or ion in the pack, but not multipliers. The basic idea of the game is that pupils have to use the cards to make chemical formulae.

The rules of Formulon are simple. The players (up to eight in any one game) are each dealt a hand of ten cards, the remainder of the pack being placed face down and the top card turned face up and placed alongside. Play then proceeds in much the same way as in rummy, each player having to do one of the following three things on his turn.

(a) Place a correct formula on the table using the cards in his hand **eg.**

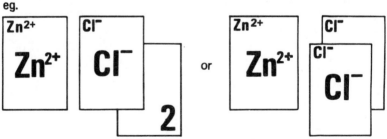

(b) Alter a formula he has placed on the table previously (but not pick it up and replace it by a completely new one)

eg.

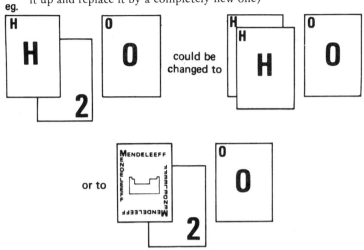

(c) Pick up a card from either the face up or the face down pile and then discard one of his cards by placing it on top of the face-up pile.

Each round of the game ends when one of the players has used up all his cards. The other players add up the valencies of all the cards they are left with, multipliers counting as their face values and Mendeleef cards as 8. The scores are noted and, after several rounds (as many as time permits), are totalled; the player with the lowest total is the winner.

Formulon is a good example of a game in which the educational content and game structure are well integrated.[55] To play effectively, players must have a basic knowledge of the properties of the different types of element and ion and of the way in which these combine to form ionic and covalent compounds. The game constitutes a vehicle for reinforcing this basic knowledge, and (through feedback from other players) of helping pupils to fill any gaps in their knowledge or fundamental misunderstandings they may have. It is also simple to organize with a class of any size, and could therefore be incorporated into any basic chemistry course at school or college level. In addition, Formulon could serve as a useful paradigm for teachers of other branches of science. Its basic principles could, for example, easily be adapted for use in a number of biological teaching situations.

CHEMSYN

Chemsyn (p 151) is a more sophisticated chemistry-based card game designed for use in teaching basic organic chemistry to senior school pupils and junior undergraduates.[77] It consists of a pack of 52 numbered cards plus a 24-page instructional booklet, each card containing detailed information about one particular organic compound. The object of the game is to convert a random distribution of cards into an ordered sequence, illustrating the way in which the various compounds can be interconverted.

Each card in the Chemsyn pack has a 'picture' side and a 'text' side; the former depicts a number of different representations of the structure of the particular compound, while the latter gives detailed information about its stereochemistry, basic properties, method(s) of preparation and reactions (see Figure 3.1).

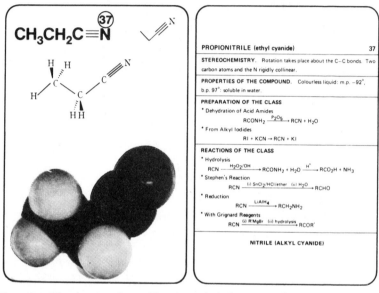

Figure 3.1: The picture and text sides of a typical Chemsyn card

Of the 52 cards in the pack, 50 represent compounds drawn from a broad range of chemical classes belonging to aliphatic and aromatic organic chemistry (eg alkanes, alkenes, alkynes, alcohols, phenols, esters, aldehydes and ketones). The number of cards devoted to each class varies from one to five and, where appropriate, both aliphatic and aromatic compounds are represented. The remaining two cards are blank and can be used to represent any compound (ie they serve as jokers).

Chemsyn can be played by a single person (solo Chemsyn) or by a group of two to five students (group Chemsyn). In each case, the basic object of the game is to build up an ordered sequence of cards that shows how the various compounds (or the chemical classes to which they belong) can be interconverted by synthesis and degradation. In solo Chemsyn, for example, the player works systematically through the pack in solitaire fashion, trying to add each successive card to the sequence he has built up. In group Chemsyn, the cards are dealt to the participants and to a central bank, and the players (operating in rummy fashion) try to build up a single common sequence by using the cards in their hands; the winner is the first person to get rid of all his cards. Part of a typical Chemsyn sequence is shown in Figure 3.2.

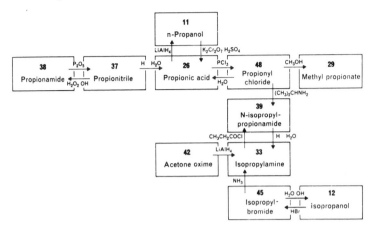

Figure 3.2: Part of a typical Chemsyn sequence showing the reagents by which each of the transformations might be achieved.

The educational philosophy behind Chemsyn is that organic chemistry is based on a relatively small number of fundamental facts and concepts, with the subject opening up in a logical manner once these have been mastered. The essential groundwork lies in a sound knowledge of functional group reactions, of molecular shape and of commonly used names and terminology for individual compounds and compound classes. The object of Chemsyn is to help students reinforce this groundwork by placing them in a challenging or competitive (and hence a motivating) situation. Although not as well integrated a game as Formulon,[55] it is a good example of how a game structure can be superimposed on an advanced learning situation in order to make it more palatable to students. Thus it could again serve as a useful paradigm for teachers of other branches of science.

Board games

The class of board games is taken to include all games which make use of a specially provided playing surface of some sort. It includes a large number of science-based games, particularly the range produced by Longmans, many of which make use of such playing surfaces. Note that the board in such a game need not be a conventional stiff board of the Monopoly type; most of the Longman's games, for example, have thin card boards that are supplied as part of the game booklets.

As a group, board games have similar characteristics to the card games discussed above, except that they are generally somewhat more complicated and, in many cases, considerably more expensive (the

Longman range are notable exceptions, since they are all produced as inexpensive A4 booklets which are dismembered and cut up in order to produce the playing materials). Like card games, their main educational uses are as reinforcement exercises, although the fact that most board games take somewhat longer to play than the average card game makes them slightly more difficult to fit into tight curricula. Again like card games, it is essential that their academic content and game structure be well integrated if they are to constitute effective educational exercises.

Board games come in a wide range of types, but it is possible to introduce a rough classification according to the way in which the playing surface is used. In the first group, of which Scrabble is a good example, the board is used simply as a matrix on which a pattern of some sort may be built up. In the second group, of which Ludo and Monopoly are typical members, the board is used to provide a predetermined linear path (or paths) along or round which players have to progress. In the third group, of which chess is probably the best-known example, the board is used as a simulated battlefield of some sort, play being both mobile and two-dimensional (rather than one-dimensional, as in the second group). In the final group, of which Shell's North Sea was one of the first examples, one-dimensional activity on the perimeter of the board is used to control two-dimensional activity on the interior. Let us now examine an example of each type of game, namely, Circuitron, The Great Blood Race, Invasion (Microbes) and The Offshore Oil Board Game.

CIRCUITRON

Circuitron (p 131) is a physics-based exercise designed for use in teaching electrical circuit theory.[30, 31] It is not a single game, but a family of five games ranging from the very simple to the very difficult. The games can be used at a wide range of levels − from upper primary, through secondary, to lower tertiary. In every game, the basic idea is the same, with players building up electrical circuits by arranging rectangular pieces representing components such as batteries, bulbs and switches on a specially designed board.

The Circuitron package consists of a board, a set of 64 cardboard playing pieces plus 64 blanks (for use in making extra playing pieces if required or replacing lost pieces), and a 32-page teacher's manual. The board consists of a rectangular matrix of 110 slots in which pieces may be placed joined by 'connecting wires' (see Figure 3.3). The various types of playing piece and the values they are assigned in the game are shown below.

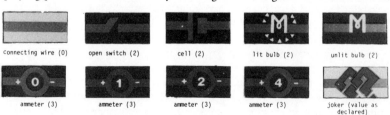

connecting wire (0)	open switch (2)	cell (2)	lit bulb (2)	unlit bulb (2)
ammeter (3)	ammeter (3)	ammeter (3)	ammeter (3)	joker (value as declared)

In a typical game, a player (or pair of players) draws a 'hand' of pieces, and then tries to place them in slots in the board in such a way as to make up a complete valid electrical circuit with as high a total score as possible (the sum of the values of the individual pieces used).

Figure 3.3: A typical Circuitron circuit (total score 19)

The five basic games that can be played using the package differ considerably in complexity and difficulty, these factors being controlled by varying the types of pieces available for selection from the pack, the number of pieces in a hand, and the detailed rules governing the formation of circuits. The package is designed for use with a group of four players, who may play as individuals or in pairs, each individual or pair trying to build up as high a score as possible over a number of rounds.

As in the case of the two card games described earlier, Circuitron is *not* intended as a substitute for either conventional teaching or laboratory work. Rather, it is designed as a reinforcement and consolidation exercise to be used immediately after new ideas have been presented or new types of circuit studied in the laboratory. For example, after pupils have just discovered by experiment that the current is constant all round a series circuit and have learned how to connect up ammeters, they could play Game 2 in which all their ammeters need to have the same reading and be joined up correctly. To help teachers make the best use of the package, the teacher's manual includes the algorithm shown in Figure 3.4 — a good example of the type of systems approach that can help to ensure that educational games and simulations are used to maximum effect.

As mentioned in Chapter 2, Circuitron was one of the first science-based games on which a thorough evaluation of educational effectiveness

WHO CAN PLAY WHAT?

How your students play the game depends on how much they know. The flow chart is to help you decide where to start. (If you are not sure how much your students know, you could give them the test on page 24)

Do they know that . . . . . . ?

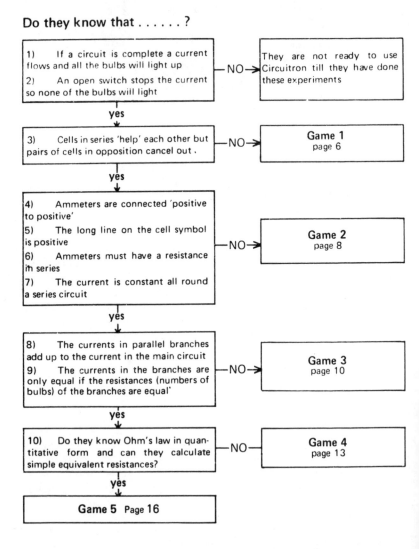

Figure 3.4: The algorithm showing how Circuitron should be used

was carried out.[30] During the pre-publication trials of the package,
111 pupils of 12 to 13 years (61 boys and 50 girls) at three comprehensive
schools in Glasgow, who had all completed Section 7 (basic electricity) of
the Scottish Integrated Science Course, were given a short objective test to
see how much they knew about circuits. They then played Circuitron for
about two double periods (roughly two and a half hours altogether), after
which they took the test again and filled in a short questionnaire on the
game. Comparison of the results of the two tests showed that a
statistically significant gain in knowledge had taken place, while the
questionnaire indicated that over half of the pupils thought they had
'learned a lot about electric circuits' from playing the game and that over
three-quarters had found it 'highly enjoyable'.

THE GREAT BLOOD RACE

The Great Blood Race (p 190) is a biology-based simulation game
designed for use in the middle and upper forms of secondary schools. It
deals with the circulation and basic physiological functions of human
blood, and forms a self-contained learning unit that can either be used on
its own or as a reinforcement for conventional face-to-face teaching.

The game (which is designed for use with groups of two to six players)
is supplied in the form of an A4 booklet whose four outer sheets (of
thin card) are used to prepare the board and other playing materials and
whose two innermost sheets (of paper) constitute an eight-page
background reader/instruction booklet for the participating pupils.
A playing set of the game is prepared by removing the four outer sheets
of the booklet, joining the first two together to form an A2 board, and
cutting up the other two to form six packs of 'chance' cards and two
'spinners'.

The way in which the game constitutes an accurate simulation of the
human circulatory system can be seen by comparing Figure 3.5 (a
schematic diagram of the circulatory system) and Figure 3.6 (which shows
the basic lay-out of the board). As its name suggests, the game consists of
a race — players having to complete a single circuit of the branching
'track'. Their rate of progress is controlled by use of a six-sided spinner,
and is helped or hindered by various chance and other factors
encountered on route.

Before they attempt to play the game, the participants are expected to
read the background/instruction booklet thoroughly to ensure they have
the knowledge needed to play effectively. As they progress through the
game, the players are confronted regularly with problems that they can
only solve if they know certain facts about human blood — a feature that
helps provide motivation for learning. This motivation is undoubtedly
increased by the fact that a player who is ignorant of certain important
properties of blood may have to withdraw from the game if he makes a
serious mistake, although the educational effectiveness of such a drastic
penalty is open to question.

The type of linear format employed in The Great Blood Race is one

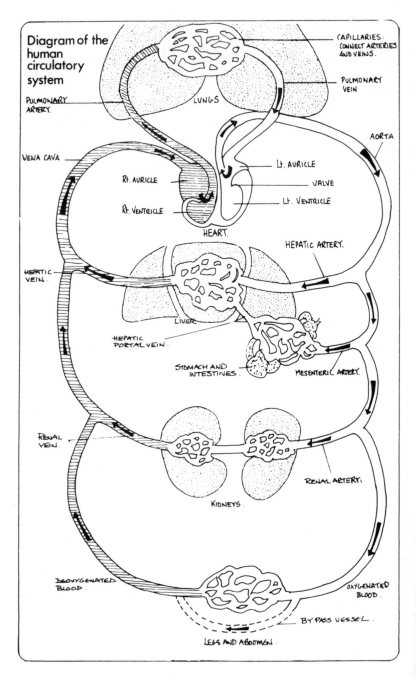

Figure 3.5: Schematic diagram of human circulatory system

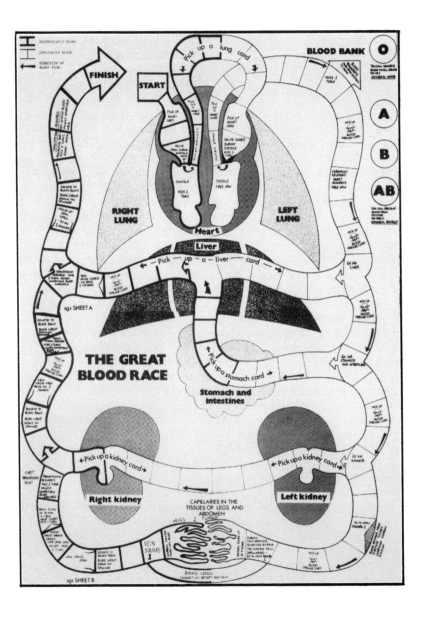

Figure 3.6: The simulation of the human circulatory system used in The Great Blood Race

that could be adapted for use in a variety of teaching situations in a wide range of subjects. In addition, use of this format (with which most pupils are familiar) generally produces an exercise whose basic rules can be picked up fairly easily; this is an important factor in determining the educational effectiveness of a game, since pupils are liable to lose interest in an exercise whose rules they find difficult or abstruse.

INVASION (MICROBES)

Like The Great Blood Race, Invasion (Microbes) (p 184) is a biology-based game designed for use at middle and upper secondary level. It simulates the 'wars' that are constantly being fought to maintain our health, and is designed to help pupils learn how it is possible for our bodies to ward off the attacks of disease-causing microbes by deploying defence mechanisms of various types. The game again constitutes a self-contained learning unit that can be used either on its own or as a reinforcement exercise.

The game, which is designed for simultaneous use by up to four pairs of players, is supplied in the form of an A4 booklet very similar to that of The Great Blood Race. The outer sections of this booklet are used to prepare the A2-sized board and other playing materials (mainly sets of 'attack', 'defence' and 'bonus' cards), while the central section again forms an eight-page background reader/instruction booklet for the participants.

The game board is divided into four independent sectors, each of which represents an organ that is under attack by pathogens (1. the skin, under attack by staphylococci, 2. the lungs, under attack by tuberculosis bacilli, 3. the sex organ, under attack by syphilis spirochetes and 4. the gut, under attack by typhoid bacilli). All four sectors have the same general lay-out, that representing the gut being shown in Figure 3.7.

Before starting the game, the participants are again expected to study the background/instruction booklet in order to acquire the detailed knowledge of the properties of the different types of pathogen and the mechanisms by which the body tries to repel them that is needed to play effectively.

The game proper takes place in two stages. In the first (the so-called 'colonization' or 'force raising' phase) the players who are attacking and defending a particular organ have to try to build up their respective forces of pathogens and defensive bodies (phagocytes and antibodies) in the areas behind the central battle zone. They do this by taking turns to draw cards from the appropriate 'attack' and 'defence' packs and trying to answer questions about their respective 'forces'; if they succeed in doing so, they win the right to colonize a certain number of hexagons and (in some cases) to collect a 'bonus' card for use in the latter stages of the subsequent battle phase. A hexagon is colonized by writing the appropriate symbol in its centre.

The second phase of the game is the actual 'battle'. Here, the players take turns to use a 'battle spinner' that tells them by how many hexagons they can advance into the central 'battle zone'. In this phase, the attacker tries to occupy as many hexagons as possible while the defender tries to

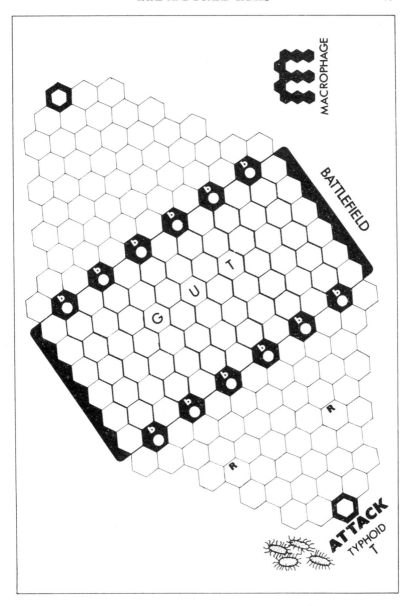

Figure 3.7: Sector 4 of the board used in Invasion (Microbes) — the gut

eliminate these pathogens by deploying phagocytes and antibodies (once both a phagocyte and an antibody have been placed adjacent to a pathogen, the latter is destroyed). Play continues until all hexagons on the battlefield have been occupied, whereupon the defender and attacker use the 'bonus' cards they have collected to destroy or reinstate further pathogens. A score sheet is then used to determine who has won, points being awarded for the different types of hexagon occupied by each player.

Invasion (Microbes) is an excellent example of the way in which a board can be used as the basis of a two-dimensional 'mobile' game and also of the way in which motivation for learning can be produced by placing students in a competitive situation in which they *have* to know certain facts before they can play effectively. (As a general rule, players who are preparing to play a game or take part in a simulation only take the trouble to acquire such background knowledge and familiarity with the rules that are essential if they are to play reasonably effectively, all other material being either skimmed through or completely ignored.) We have seen that one of the main factors in determining the overall effectiveness of an educational game is the extent to which the academic content and game structure are integrated. Invasion (Microbes) demonstrates one method by which such integration can be achieved.

THE OFFSHORE OIL BOARD GAME

The Offshore Oil Board Game (p 196) began life as North Sea, a family board game based on Britain's North Sea oil industry that was developed by Shell (UK) Ltd during the mid-1970s.[78] The game was subsequently converted into a decision-making exercise for use in the Association for Science Education's Science in Society course, the original expensively produced boxed game being re-designed as an inexpensive 'do-it-yourself' kit that could be made up by teachers.[79]

A playing set of the game (which is designed for use by up to five people — four players and a 'banker') consists of a board, a set of playing materials (playing pieces, money, chance factor cards, etc), five student booklets, five student cards and a teacher's guide. The student booklets contain detailed background information about the offshore oil industry, together with the full rules of the game, and are intended to be studied at home before the game starts. The student cards carry a diagram of the board on one side (see Figure 3.8). This is meant to be studied at home in conjunction with the rules. On the other side, they carry an easy-to-follow synopsis of the game (see Figure 3.9) designed for use during actual play.

The Offshore Oil Board Game is a realistic simulation of the actual process by which North Sea oilfields are discovered and subsequently brought into production. It begins with an auction of all the concessions in one area of the board, each player having to acquire at least one if he is to take any further part in the game (this gets the exercise off to an exciting, highly competitive start). Next, players have to go through the process of exploring for oil, appraising any field(s) they discover, developing one of their fields, bringing it into production and using the

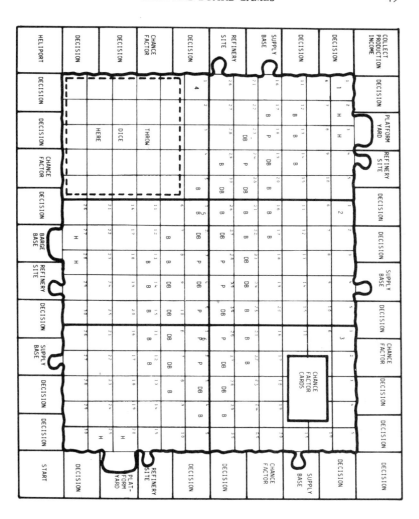

Figure 3.8: The lay-out of The Offshore Oil Board Game board

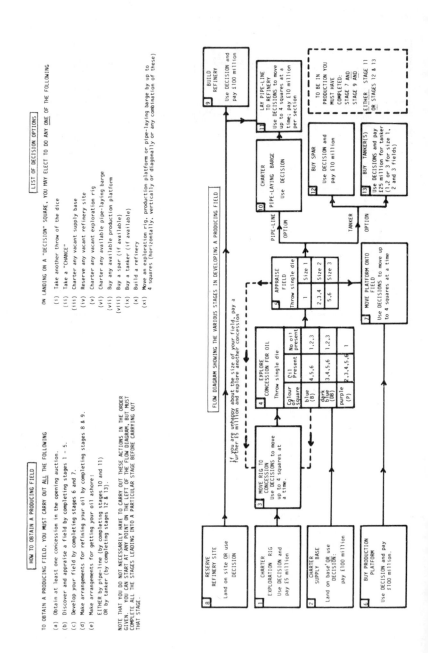

Figure 3.9: The synopsis of The Offshore Oil Board Game given on the student card

resulting income to pay off the loans needed to finance earlier operations. The winner is the first player to bring a field into production and pay off all his loans.

A key feature in the design of the game is that activity on the interior of the board (which represents a potential oil-bearing sedimentary basin in the North Sea) is controlled by activity on the perimeter (which represents the surrounding coastline). Players move round the perimeter in Monopoly fashion by throwing two dice, making use of the 'decision' squares to carry out their programmes of activities on the interior (see Figure 3.8).

The educational aims of The Offshore Oil Board Game are twofold, namely, to teach the participants about the way in which the offshore oil industry operates and to help them develop the ability to think rapidly and effectively under pressure.[79] The key to achieving both these aims is the use that is made of the 'decision' squares on the perimeter of the board. Every time a player lands on one of these squares, he has to decide (on the basis of his knowledge of how offshore oilfields are developed and his appraisal of the current tactical state of the game) what his most effective next step would be. The idea of using 'decision' squares in a game, and the associated idea of using one-dimensional activity on the perimeter of a board to control two-dimensional activity on the interior, could well be adapted for use in other teaching situations.

Chapter 4

Other manual exercises

Strictly speaking, the term 'manual' applies to all games, simulations and case studies that do not involve use of the computer — including card and board games of the type described in the last chapter. For the purposes of this chapter, however, we shall consider the term as applying only to those non-computer-based exercises where the main resource material consists of sheets of paper, leaflets or booklets rather than cards or a board. This has traditionally been the most common medium in which academic games, simulations and case studies have been written.

The most important characteristics of the manual medium (using the term in the limited sense defined above) are its great versatility and flexibility, and the comparative ease with which resource materials can be produced and duplicated. The former makes it possible for manual exercises to achieve a much wider range of educational objectives than card or board games (which, as we have seen, are largely limited to low-level cognitive objectives) and allows them to be used in a much greater variety of roles, covering all the areas outlined in the first part of Chapter 2. The latter means that anyone with access to a typewriter and a photocopier can produce all his/her own materials and run off as many copies as he/she needs easily and inexpensively. This again contrasts sharply with the card and board game media, in which the resource materials are generally more difficult both to produce and to duplicate; boards, for example, have either to be laboriously manufactured by hand or printed by some relatively expensive process.

Despite the wide range of formats in which manual exercises are produced and the wide range of uses to which they are put, they can usefully be divided into four basic types according to their underlying structure, namely *linear, radial, composite* and *multi-project* exercises.[80] The main characteristics of each of these classes will now be discussed, and specific illustrative examples (that could serve as useful paradigms for similar developments in other subject areas) examined.

Linear exercises

The two fundamental structures on which virtually all manual exercises are based are the so-called linear and radial structures, each of which can be used for the achievement of distinct (although not necessarily mutually exclusive) sets of broad educational aims and objectives.[80] The essential structural characteristics of exercises that have a *linear* format are as follows:

— the participants progress systematically through a predetermined series of activities;
— all participants have the same basic resource materials and carry out the same basic set of activities.

In common with all other types of game, simulation and case study, linear exercises can be used to achieve a wide range of educational objectives related to their subject content. In addition, they have the following specific educational characteristics:

— the progressive nature of their structure enables a complicated case study to be broken down into easily manageable stages and clearly illustrates the relationship of each part to the whole;
— they can be used to foster the development of a wide range of skills, including problem solving, analytical and decision-making skills.

These characteristics will now be illustrated by examining two specific exercises, namely Project 1 from the Central Heating Game and What Happens When The Gas Runs Out?

PROJECT 1 FROM THE CENTRAL HEATING GAME

The Central Heating Game (p 130) is in fact a multi-project exercise[76] (a collection of five projects all based on the theme of domestic central heating), and will be described later in this chapter. Its first project is, however, an excellent example of the type of linear exercise under discussion, and will therefore be examined now. This is designed for use in the teaching of physics, engineering and architecture at upper secondary and lower tertiary level, and takes the form of a highly structured simulated case study on the heating requirements of a typical three-bedroomed bungalow. The project takes roughly two hours to complete and can be used with a class of up to 24.

The structure of the project is shown schematically in Figure 4.1. Before the project starts, the members of the class are each issued with a copy of an introductory booklet to study at home. This booklet contains general information about domestic central heating and home insulation, and is designed to give the participants the background knowledge that they need to carry out the work of the project (it is used as an introduction to all the projects in the Central Heating Game package).

At the start of the project, each student or pair of students is issued with a data sheet giving basic architectural and technical information about the bungalow to be studied together with a copy of project sheet 1(a). This provides them with the theory that they need to carry out Stage 1 of the case study — the calculation of U-values (thermal transmittance coefficients) of some of the structural components of the bungalow (the walls and windows in this case). The sheet also gives them detailed instructions on how to calculate these U-values together with a partly completed table in which the answers to the various stages of the calculations may be inserted.

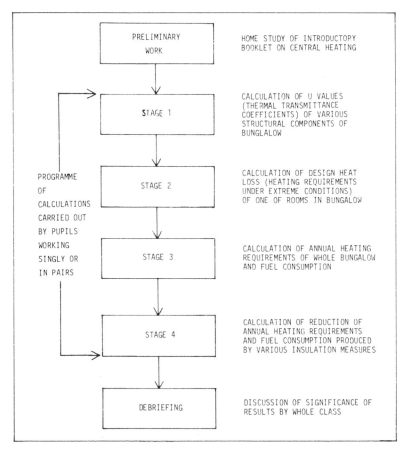

| PRELIMINARY WORK | HOME STUDY OF INTRODUCTORY BOOKLET ON CENTRAL HEATING |

Figure 4.1: Schematic structure of Project 1 from Central Heating multi-project pack

As soon as Stage 1 has been satisfactorily completed, the students are issued with project sheet 1(b). The first part of this sheet contains a table giving the correct answers to the U-value calculations carried out in Stage 1, plus the U-values of all the other structural components of the bungalow, so that the students can check their earlier work. It also provides them with the theory and instructions they need to carry out Stage 2 of the case study (calculation of the design heat loss of one of the rooms of the bungalow — the kitchen) and a table in which the answers to the various calculations may be inserted.

As soon as Stage 2 has been satisfactorily completed, the students are issued with project sheet 1(c). This again contains a table that gives the correct answers to the calculations carried out in the previous stage, plus the corresponding design heat loss figures for all the other rooms of the bungalow. It also provides the theory and instructions needed to carry out Stage 3 of the case study (calculation of the total annual heat energy requirements of the bungalow) and a table in which to fill in the answers to the various calculations.

On completion of Stage 3, the students are issued with the final project sheet, which again provides the correct answers to the calculations just carried out. It also contains detailed instructions on how to carry out the final stage of the case study — calculation of the effects on the annual heat requirements of the bungalow of introducing various additional insulation measures (loft insulation, cavity wall insulation and double glazing) — together with tables in which the answers to the various calculations may be inserted.

Once Stage 4 has been completed, the students are issued with a second data sheet that contains the answers to the final set of calculations. The project is then brought to a conclusion by holding a short debriefing session in which the significance of the results of the project are discussed by the whole class. Such debriefing sessions are a vital part of educational games, simulations and case studies of all types.

The method of approach employed in the above project enables a highly complex and challenging problem to be broken down into easily manageable stages, and also gives the students a clear understanding of the significance of the various calculations involved. The linear format is ideally suited to this type of use.

WHAT HAPPENS WHEN THE GAS RUNS OUT?

What Happens When The Gas Runs Out? (p 177) is a chemistry-based interactive case study designed for use with junior science undergraduates and senior secondary pupils of roughly A-level standard.[10, 43] Its main aims are to provide a vehicle for educating 'through' science (see Chapter 2) and to help demonstrate the social and economic implications of science and technology. The exercise involves appraising Britain's likely reserves of natural gas and formulating a policy for providing a viable replacement when these reserves are exhausted. The project is designed to fit into a standard three-hour laboratory session, the participants working in co-operative groups of four to eight.

The structure of the exercise is shown schematically in Figure 4.2. It begins with a brief introduction by the teacher or lecturer in charge. The students are then divided into their working groups and issued with copies of a sheet that describes the work to be carried out in Part 1 of the exercise and provides them with all the necessary background information and data. Part 1 involves carrying out a quantitative examination of the calorific properties of manufactured town gas and natural gas, and discussing the sort of problems likely to be encountered in converting gas appliances from one to the other. This stage of the exercise takes roughly 40 minutes.

Once Part 1 has been satisfactorily completed, the students are issued with a second sheet describing the work to be carried out in Part 2 and providing all the necessary information and data. Part 2 involves relating the quantity of natural gas estimated to be present in the North Sea to

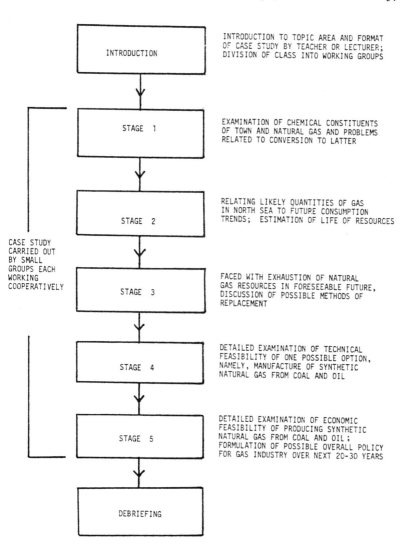

Figure 4.2: Schematic structure of What Happens When The Gas Runs Out?

likely future consumption trends and hence estimating how long the gas is likely to last. It again takes roughly 40 minutes.

On completion of Part 2, the students are issued with a further work sheet for use in Part 3 of the exercise. This takes as its starting point the fact that Britain's natural gas reserves are almost certain to become exhausted well within the lifetime of the students (the conclusion that they should have reached in Part 2), and involves discussing possible alternative sources of supply and appraising their relative advantages and disadvantages. This stage takes roughly 20 minutes.

The next two stages of the exercise (Parts 4 and 5) consist of a detailed appraisal of one of the most promising options that should have emerged from the discussions in Part 3, namely manufacture of synthetic natural gas from coal or oil. Part 4 (35 minutes) is devoted to an examination of the technical feasibility of this particular option, and Part 5 (45 minutes) to an appraisal of its economic viability and formulation of a possible overall policy for Britain's gas industry over the next 30 years. The students are again issued with a new work sheet at the start of each stage.

What Happens When The Gas Runs Out? is essentially a linear programme in which the participants are led through a series of activities whose objectives are initially mainly in the lower part of Bloom's cognitive domain[11] and finish up in the higher regions. Such a progression is a feature of many exercises of this type, and is something that the authors believe can be achieved by employing a linear structure of the type shown; this structure enables each stage to build systematically on the work of its predecessors, thus allowing a progression from simple to complex ideas to take place easily and naturally. Like Project 1 from the Central Heating Game, the exercise also provides a good example of the effective use of resource materials in a linear teaching situation.

Radial exercises

The second fundamental structure that can be used in manual exercises is the so-called *radial structure*. Exercises designed in this format have the following basic structural characteristics:

— each participant (or group of participants) carries out a set of activities specific to a given role in a scenario or particular point of view in a problem situation and then presents information or argues a case at a plenary session or simulated meeting;
— the various participants (or groups of participants) have different resource materials and carry out different (albeit often related) activities.

Like linear exercises, radial exercises can be used to achieve a wide range of educational objectives related to their subject content, but also have a number of educational characteristics that are specifically related to their structure:

— they enable the different arguments or points of view in a complicated problem to be identified, examined in detail, and subjected to informed criticism and discussion;
— they foster the development of a wide range of useful skills (particularly those related to the preparation, presentation and defence of arguments), and can also be used to develop desirable attitudinal traits (such as a willingness to listen to the points of view of other people or an appreciation that problems can generally be viewed in a number of different ways).

These characteristics will now be illustrated by taking a detailed look at three typical radial exercises, namely Fluoridation?, The Amsyn Problem and Proteins as Human Food.

FLUORIDATION?

Fluoridation? (p 194) is a role-playing simulation game designed for use as a case study and 'mind-broadening' exercise in science in society, modern studies and health education courses at both upper secondary and tertiary level.[81] It is based on the hypothesis that an area health authority (for the imaginary Hadley area) is considering the principle of fluoridation of the public water supply, and takes the form of a simulated public meeting called by one of the Community Health Councils to discuss the question. The exercise takes between one and a quarter and two hours to complete (depending on the numbers involved and on the level of sophistication of the participants) and can be used with a class of between 13 and 24 students.

The structure of Fluoridation? is shown in schematic form in Figure 4.3. Roughly a week before the exercise is due to take place, each participant is given a copy of an introductory booklet to read at home. This contains background information about those aspects of local government structure relevant to the game, a summary of the main features of the simulated area, and a review of the format and structure of the exercise. At this stage, the participants are also allocated their roles and given the appropriate briefing booklet so that they can prepare the arguments they are to present.

The game itself takes the form of a structured debate in which representatives of the various groups that support and oppose fluoridation present their respective cases to the members of the Community Health Council, who have the task of deciding whether or not to support fluoridation when the issue is discussed at a higher level. The exercise has been designed in such a way that the main arguments commonly presented in favour of fluoridation are shared between the various supporters, while those generally raised by the pressure groups that oppose fluoridation are shared between the objectors (see Figure 4.3). The chairman of the Council controls the debate, calling the various speakers in a predetermined order and using any time that remains for an open discussion.

Fluoridation? is a typical radial exercise in that it (a) provides the participants with the basic facts regarding the issue being examined, and (b) shows that these basic facts can be looked at from more than one point of view. In particular, it highlights the type of conflict that almost invariably arises between the protagonists of a controversial measure, who generally produce detailed arguments to show that it would be technically or economically beneficial to the community as a whole, and its opponents, who generally claim that its introduction would violate the rights of the individual or produce unacceptable (albeit often unquantifiable) environmental or social side-effects. Such issues, whose resolution has nearly always to be based on the formulation of value

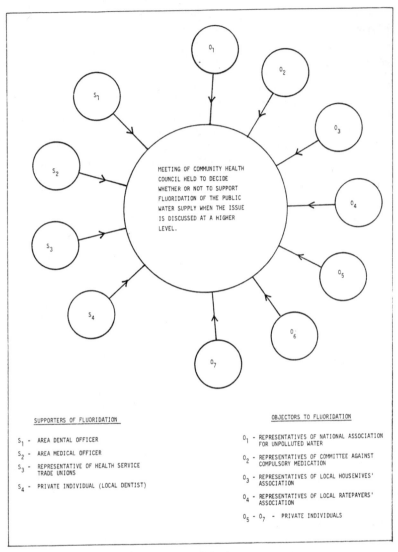

MEETING OF COMMUNITY HEALTH COUNCIL HELD TO DECIDE WHETHER OR NOT TO SUPPORT FLUORIDATION OF THE PUBLIC WATER SUPPLY WHEN THE ISSUE IS DISCUSSED AT A HIGHER LEVEL.

SUPPORTERS OF FLUORIDATION

S_1 – AREA DENTAL OFFICER
S_2 – AREA MEDICAL OFFICER
S_3 – REPRESENTATIVE OF HEALTH SERVICE TRADE UNIONS
S_4 – PRIVATE INDIVIDUAL (LOCAL DENTIST)

OBJECTORS TO FLUORIDATION

O_1 – REPRESENTATIVES OF NATIONAL ASSOCIATION FOR UNPOLLUTED WATER
O_2 – REPRESENTATIVES OF COMMITTEE AGAINST COMPULSORY MEDICATION
O_3 – REPRESENTATIVES OF LOCAL HOUSEWIVES' ASSOCIATION
O_4 – REPRESENTATIVES OF LOCAL RATEPAYERS' ASSOCIATION
O_5 – O_7 – PRIVATE INDIVIDUALS

Figure 4.3: Schematic structure of Fluoridation?

judgements rather than on the rational appraisal of facts, are ideally suited to treatment using a radial rather than a linear format. Fluoridation? provides a good example of the type of approach that may be employed in dealing with issues of this kind in a classroom.

During its final pre-publication field trials in three Aberdeen schools, an attempt was made to evaluate the educational effectiveness of Fluoridation?[82] This proved extremely encouraging, indicating that the exercise was succeeding in achieving both its cognitive and its affective objectives and clearly demonstrating that it was highly popular with both pupils and teachers. One interesting (and not altogether unexpected) outcome of this evaluation was that the participants tended to become polarized by their roles, those who were given roles supporting fluoridation showing a distinct positive attitude shift towards fluoridation as a result of playing the game, and those given opposing roles showing an attitude shift in the opposite direction. Such polarization is something that must be kept under careful review by designers and users alike, since its effects can, in certain circumstances, be counter-productive or even harmful.

THE AMSYN PROBLEM

The Amsyn Problem (p 173) is a role-playing simulation game designed for use with senior secondary pupils and junior undergraduates as a vehicle for educating 'through' science and for demonstrating the social relevance of chemistry.[43] It is based on a typical industrial problem situation, in which a small chemical firm have to decide how they can reduce the impurity levels in their effluent in order to meet new regulations. The exercise is designed for use with an optimum number of 16 participants and takes roughly two to three hours to complete.

The game scenario is based on the hypothetical firm of Amsyn Ltd, whose main activity is the manufacture of aromatic amines. Their current production process produces large quantities of heavily polluted effluent, however, which has until now been discharged directly into a nearby river — with disastrous environmental results. In an attempt to clean up the river, the local district council plan to introduce stringent regulations regarding the impurity levels of industrial effluent, and Amsyn will either have to devise a means of satisfying these regulations or face partial or total closure of their plant — something that would be socially unacceptable in an already depressed area.

It is against this background that the management of Amsyn Ltd have called a meeting to discuss the situation with the other interested parties, namely the trade union representatives of the workforce and representatives of the district council. Each party, although wishing to find a mutually satisfactory solution, naturally views the problem with a different set of values, priorities and responsibilities (see Figure 4.4).

The exercise starts with a short tape/slide programme, which introduces the participants to the game scenario. The class is then divided into three groups (management, trade union representatives and district council representatives) and each group is given copies of a booklet that defines its role and provides its members with all the information they require. The groups study their respective booklets, and each then tries to decide on the course of action it would prefer.

When the participants re-assemble at a management-chaired meeting, delegated spokesmen from the three groups present the various proposed solutions to the problem, after which a general discussion takes place in order to see if a mutually satisfactory compromise can be reached. If, at

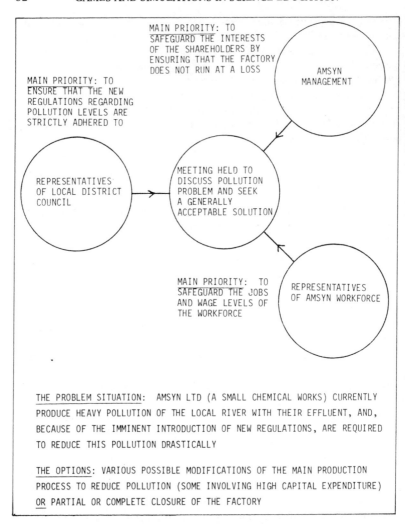

Figure 4.4: Schematic structure of The Amsyn Problem

the end of the allotted time, no agreement has been reached, the management is required to propose a definite plan, and the other parties are asked for their reactions.

Like Fluoridation?, The Amsyn Problem allows a complex issue, involving a number of seemingly incompatible technical, economic and social issues, to be examined from the points of view of the various interested parties. In this case, however, the object of the exercise is to produce a generally acceptable compromise solution to a multi-sided problem rather than to reach a clear-cut decision on a straightforward 'either-or' issue; such situations also lend themselves to treatment using a radial approach.

Another valuable feature of the exercise is that it places the intrinsic scientific content in the context of a highly realistic scenario. Thus, the chemistry is seen to be related to a variety of other disciplines, and it is shown that what the students might at first suppose to be purely technical decisions are often influenced by social, economic and even moral considerations. As a result, the basic chemical equations are seen to have relevance and implications far beyond the textbook, or even the laboratory.

PROTEINS AS HUMAN FOOD

Proteins as Human Food (p 168) is an interactive case study designed specifically to help science students at upper secondary and lower tertiary level develop communication, inter-personal and decision-making skills.[10,42] It deals with the general topic of proteins and the world food shortage and, although it has its main basis in chemistry, also involves a number of other disciplines, including biology, nutritional science, economics and geography. The exercise is designed for use by groups of six students and takes roughly one and a half to two hours to complete.

A problem commonly encountered in group tutorials (even when a skilful tutor is involved and a mini-lecture is avoided) is that some students, possibly through lack of confidence or knowledge, fail to become involved in the discussion, which is often dominated by the tutor and one or two group members. Thus, the central problem facing the designer of a communication exercise is how to make sure that all the participants are involved to an approximately equal extent.

In Proteins as Human Food, the approach adopted is to have a group leader whose only task is to organize and control the discussion and five other members each of whom is initially in sole possession of one component of the information needed by the group if it is to hold a meaningful discussion on the world protein problem (see Figure 4.5). Each of the six members of the group is given an individual booklet that (a) tells him what will happen in the exercise, (b) defines his role in the discussion, and (c) provides him with the detailed information specific to his role. In this way, each member *has* to contribute to the discussion, thus helping the more diffident overcome their reluctance to speak.

The effectiveness with which Proteins as Human Food achieves its basic design objectives has been determined by means of a systematic evaluation involving seven groups (42 students) in three separate educational establishments (science undergraduates at the University of Glasgow, mature education students at the University of Liverpool and sixth-formers from a large Scottish comprehensive school).[10,42] This took the form of a written assessment involving the completion of pre- and post-tests by all the students who took part in the exercise, plus an analysis of tape recordings of several of the discussion sessions. Although the sample size was too small for rigorous statistical analysis, the written assessment clearly indicated that the exercise was having some success in achieving its various aims, while the analysis of the tape recordings was even more

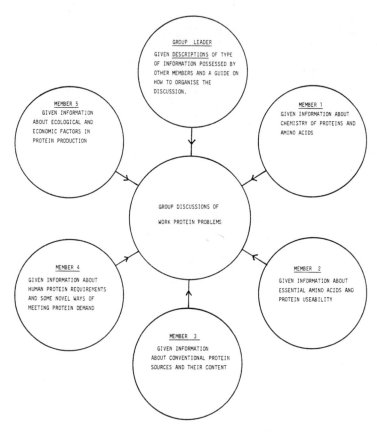

Figure 4.5: Schematic structure of Proteins as Human Food

encouraging. These demonstrated quite clearly that all six members of each of the groups studied had made a significant contribution to the discussion, with no individual member spending less than 10 per cent of the total time communicating.

Clearly, Proteins as Human Food could serve as a useful model for the development of similar communication exercises in other subject areas.

Composite exercises

Although many manual exercises have relatively straightforward structures that can be classified simply as 'linear' or 'radial', many others have rather more complicated structures, and cannot be so easily classified. Such exercises can, however, generally be broken down into a combination of linear and radial elements, and can therefore be said to have *composite structures*. By employing a composite structure, it is possible to produce an exercise that has the educational advantages associated with both linear and radial structures.

Exercises of this type are generally more complicated than those described so far, and their characteristics are probably best illustrated by taking a look at specific examples. Three such exercises will be examined. The first (Polywater) is essentially a linear exercise with some radial elements built in. The second (Power for Elaskay) is the opposite, namely a radial exercise with some linear elements built in. The third (The Power Station Game) is considerably more complicated, and cannot be so easily described (it consists of three parallel linear structures, each with some radial elements built in, leading into a radial structure which itself gives rise to a further, completely separate radial structure — see Figure 4.8).

POLYWATER

Polywater (p 167) is an interactive case study designed to help senior chemistry undergraduates develop library and communication skills.[10, 43] It is based on the controversy over the existence or otherwise of 'polywater' that raged in scientific literature between 1966 and 1973. (Polywater was believed to be a polymeric form of water created when steam was condensed in small capillaries. After seven years of intensive study and debate — and the publication of over 500 scientific papers — it was realized that the 'phenomenon' was a result of impurities partly produced by leeching of silica from the glass.) The full exercise takes roughly five or six hours to complete, although a shortened version can be fitted into a standard three-hour laboratory period if time is restricted. It is designed for use by groups of six to eight students, each with its own tutor.

The structure of Polywater is shown schematically in Figure 4.6, which clearly shows how the radial elements are built into what is essentially a linear sequence of activities. The exercise begins with a brief introduction by the tutor and the study of a short introductory booklet by the students. Each student in the group is then issued with a card that gives one or more references to early papers on the polywater phenomenon. The students are told to look these up in the library, read them, and prepare short precis of their contents. After about 45 minutes, the group re-assembles, and each student gives a short report on his findings. The group then discusses the situation revealed by the early papers, and discusses what further work could usefully be done on the subject.

Once the discussion has run its course, each student in the group is given one or more further references to look up, read and precis — this time on papers published when work on polywater was at its height. The students again report back to the rest of the group, which then holds a further discussion of the current state of the research and suggests what further lines might be pursued.

After completion of this second discussion session, the students are given a final set of references to study, this time on papers published when the true nature of the 'polywater phenomenon' was being realized. This is followed by a third feedback and discussion session, after which the exercise is brought to a conclusion with a short debriefing.

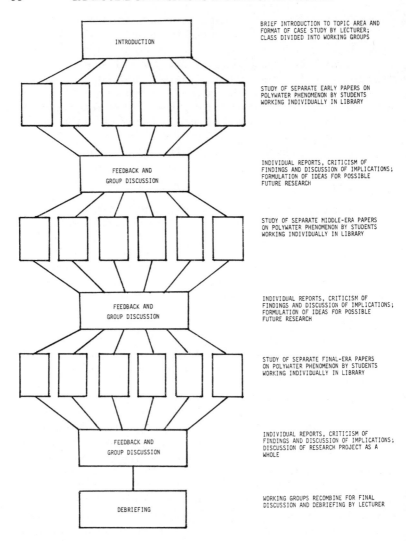

Figure 4.6: Schematic structure of Polywater

Polywater has a number of valuable educational characteristics, serving not only as a case study on an extremely interesting area of chemical research and as a vehicle for developing useful library and communication skills, but also as a forceful demonstration that not everything that is published in learned journals is necessarily correct; it thus shows that the attitude of a research worker should be one of healthy scepticism rather than blind belief in the written word.

POWER FOR ELASKAY

Power for Elaskay (p 139) is a simulated case study on alternative energy sources.[70, 71] It has its main basis in physics, but also draws upon a number of other disciplines, including economics and geography. The case study is designed as an integral part of a self-contained structured lesson on alternative energy that can be incorporated into a variety of courses at upper secondary and lower tertiary level, including science in society, general studies, physics and engineering courses. It is suitable for use with a class of up to 25, and takes roughly two to three hours to complete.

The structure of Power for Elaskay is shown schematically in Figure 4.7. The problem facing its designers was that of:

(a) giving all members of a class a general appreciation of the basic technical principles underlying the various alternative sources of energy currently available;

(b) enabling the class to carry out a detailed appraisal of the technical and economic feasibility of exploiting each of these various sources of energy;

(c) completing the work in roughly three hours.

Since (c) clearly precluded detailed study of all the various sources of energy by all members of the class, it was decided to adopt the approach outlined in Figure 4.7, and to combine an introductory lesson on the basic principles of alternative energy with a simulated case study carried out by the class. The case study involves developing a 50-year rolling programme for meeting the electricity requirements of the hypothetical island of Elaskay (supposedly located somewhere off the west coast of Scotland) by exploiting the island's natural energy resources (peat, solar energy, wind energy, tidal energy and hydro-electric power). It combines detailed systematic study of each of the five possible resources by small working groups (the linear element of the exercise) with a plenary session in which the various groups report their findings and the class uses these to develop a viable rolling programme (the radial element). Each member of the class is issued with an introductory sheet describing the scenario and structure of the case study, together with resource material specific to the particular form of alternative energy he is to examine (a project sheet plus a work sheet in which to insert his findings).

Power for Elaskay has a number of interesting educational features that could again be adapted for use in other teaching situations. Its use of five parallel case studies feeding into a plenary session allows the subject area to be studied both in depth and in breadth, and helps to foster the development of a wide range of useful skills, including analytical, decision-making, inter-personal, communication and debating skills. The various teachers and lecturers who carried out field trials of the exercise found the peer teaching that takes place in the plenary session particularly effective in helping students to gain confidence and competence in public speaking.

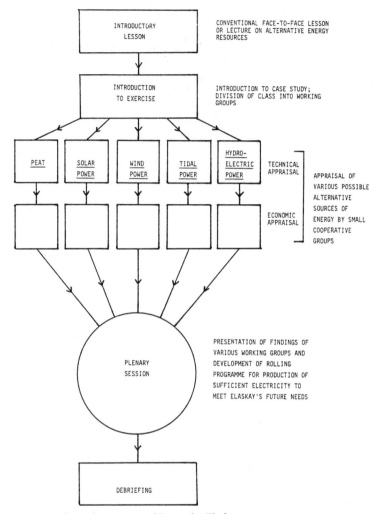

Figure 4.7: Schematic structure of Power for Elaskay

THE POWER STATION GAME

The Power Station Game (p 144) is a large-scale simulation game that was originally developed as a 'mind-broadening' exercise for physics students of roughly A-level standard. It is based on the hypothesis that a decision has been reached to build a new 2000 MW power station in a certain (imaginary) area, the object of the game being to decide what type of station to build (coal, oil or nuclear) and where to site it. The participants (optimum number 18) are divided into three competing groups, each of which has to prepare as strong a case as possible for building one particular type of station and draw up detailed proposals for their scheme. The three

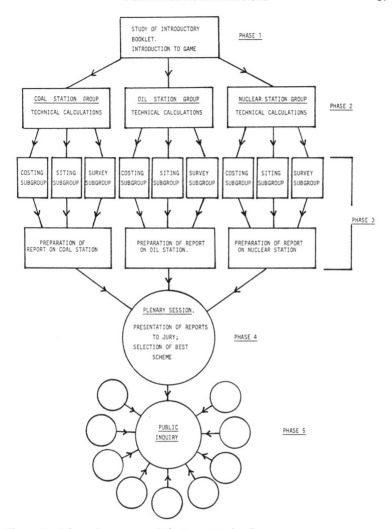

Figure 4.8: Schematic structure of The Power Station Game

teams then present their cases at a plenary session at which an independent jury decides which scheme should be adopted, after which the game is completed by holding a simulated public inquiry into the chosen scheme. In all, the exercise takes two and a half days to complete.

The structure of The Power Station Game is shown schematically in Figure 4.8, which clearly demonstrates the composite nature of the exercise. The game falls into five distinct phases, each of which is designed to achieve a different set of educational aims and objectives. In the first phase (which precedes the game proper), the participants are given an introductory booklet to study at home. This contains a general description of the electricity generation process, and is designed to give them the background knowledge needed to play the game effectively.

In the second phase of the game, the students (now divided into three competing groups) have to carry out a series of technical calculations on their particular station. These are designed to show the relevance of physics to an important real-life situation, to provide experience of the interpretation of realistic data and to give them a feel for handling large numbers. They involve working out the energy losses at each stage of the generation process, calculating the fuel and cooling water requirements of the stations, and determining the rates at which waste products are produced. The calculations take roughly three hours to complete and constitute the main scientific content of the game.

In the third phase of the game, the three groups have to prepare the cases for their respective stations. This involves calculating the capital and running costs, choosing the most suitable site and lay-out, and examining the cases likely to be made for the two rival stations with a view to identifying possible weaknesses. These three tasks are carried out by working sub-groups, which then re-combine so that the groups can draw up their final proposals and decide how best to present them at the plenary session. This phase of the game is designed to achieve a wide range of educational objectives, both cognitive and affective, including the cultivation of decision-making and inter-personal skills and the demonstration of the need for effective cross-disciplinary co-operation in tackling a complicated problem like the design of a power station.

The fourth phase of the game consists of the plenary session at which the three groups present and defend their cases. Here, the participants can develop their public-speaking and debating skills, and also learn to appreciate that real-life problems seldom have clear-cut solutions (the game is tuned in such a way that equally strong cases can be made out for all three types of station).

The final phase of The Power Station Game consists of a simulated public inquiry at which representatives of the team that had their scheme adopted have to defend it against the various objections inevitably raised when any major industrial development is proposed. It is also designed to help the participants develop their public-speaking and debating skills, and make them aware of the large number of social, environmental, amenity and other factors which must be taken into consideration before a final decision can be reached regarding a project like the construction of a power station. In addition, it is designed to make them realize that a given situation can be viewed in a number of ways, and thus make them readier to appreciate the points of view of other people.

Although primarily a physics-based exercise, the multi-disciplinary nature of The Power Station Game makes it suitable for use in a wide range of educational situations. Since its publication in 1976, it has been used as a case study with science and engineering students in schools, colleges and universities in all parts of the world,[53, 61] and has also been extensively used with non-science students as a vehicle for demonstrating the social relevance of science and technology[54] (this can be done by missing out the technical calculations). A simplified version of the game has also been developed for use in the Association for Science Education's Science in Society course for schools (see p 140). In addition, The Power Station Game has stimulated the development of a large number of further science- and engineering-based exercises.[53, 62] For example, the great popularity of its public inquiry phase during field trials prompted three members of the team who developed the game to produce a separate exercise based solely on such a simulated public inquiry;[16, 83] this has itself since been published in two different formats (see pp 166 and 169), and has proved extremely successful.

Multi-project exercises

A second group of manual exercises that cannot be classified simply as linear or radial are those containing more than one project. The individual exercises that make up such multi-project packs can, however, generally be assigned to one or other of these classes, although some belong to the class of composite exercises just examined.

Multi-project exercises have two basic characteristics that make them useful from an educational point of view. First, they allow a single set of resource materials to be used in a variety of different teaching situations, the teacher or lecturer being able to select the project most suited to his particular needs. Second, the multi-disciplinary nature of many exercises of this type often allows them to be used in the teaching of more than one subject. These features will now be illustrated by examining two specific multi-project exercises, namely the Central Heating Game and Hydropower.

CENTRAL HEATING GAME

The Central Heating Game (p 130) is a multi-disciplinary multi-project pack based on the general theme of domestic central heating and insulation.[76] Although the pack has its basis in physics, the five projects included in the package can also be used in the teaching of a wide range of other subjects (including architecture, economics, engineering, home economics, general studies and science in society) at both upper secondary and tertiary levels. Each of the five projects takes roughly two hours to complete.

The Central Heating Game is a self-contained package containing all the resource materials needed to use it with a class of up to 24 people plus a comprehensive teacher's guide giving full instructions on how to run the various projects. The resource material consists of the following:

— 24 copies of an introductory booklet entitled 'General Information about Central Heating'. This serves as an introduction to all five projects;
— 12 copies of Bungalow Sheet 1 (which gives detailed architectural and structural information about the bungalow that is used as a

basis for the various projects);
— 12 copies of Bungalow Sheet 2 (which gives detailed technical data
   on the various central heating and insulation systems that can be
   used in the bungalow);
— 12 copies of Bungalow Sheet 3 (which gives detailed economic data
   on these central heating systems and insulation methods);
— 12 copies each of Project Sheets 1(a), 1(b), 1(c) and 1(d) (which
   serve as work sheets for the various stages of Project 1);
— 12 copies each of Project Sheets 2(a) and 2(b) (which serve as
   work sheets for the two stages of Project 2);
— 6 copies of Project Sheet 3;
— 2 copies of Project Sheet 4;
— 3 copies of Project Sheet 5.

The logistical principle behind the Central Heating Game is the use of
a single set of resource materials (the introductory booklet and three
bungalow sheets) as the basis of five different projects, each of which deals
with the general theme of domestic central heating, but enables a different
aspect of the subject to be examined. The way in which this is done will
now be illustrated by describing the five projects briefly.

Project 1. This project (which has already been examined in detail in
the section on linear exercises) is a progressive technical case study that
effectively involves using the data in Bungalow Sheet 1 to derive that in
Bungalow Sheet 2. It is designed for use in the teaching of physics,
engineering, architecture, home economics and science in society.

Project 2. This is also a progressive case study, dealing with the
economic aspects of central heating and insulation; it involves using the
data in Bungalow Sheet 2 to derive that in Bungalow Sheet 3. The project
is designed for use in the teaching of economics, general studies and
science in society. As in Project 1, the class works singly or in pairs.

Project 3. This is an evaluative case study that involves using the
information and data given in the introductory booklet and bungalow
sheets as a basis for (a) choosing the most suitable central heating system
for the house and (b) appraising the cost effectiveness of the different
insulation measures. It is designed for use in the teaching of home
economics, economics, general studies, architecture and science in society.
The students work in small co-operative groups, each containing four to
six people.

Project 4. This is a role-playing exercise simulating consumer research
carried out by two separate groups, each comprising half the class. Each
group is issued with copies of all the resource material, and has the task
of questioning a 'householder' (assumed to be the owner of the bungalow)
in order to determine the central heating system best suited to his or her
particular requirements. The project is designed for use in the teaching of
home economics, economics, general studies and science in society.

Project 5. This is a simulation game in which three competing groups
(each comprising one third of the class) try to persuade the managing
director of a building firm to use a central heating system based on their

particular fuel (solid fuel, oil or electricity) in a new housing estate. The three groups are again given copies of all the resource material. The project is designed for use in the teaching of home economics, economics, architecture, general studies and science in society.

The above descriptions illustrate the versatility and flexibility of the multi-project pack format. Suppose, for example, that a teacher wished to use the pack in a science in society-type course. Such a teacher would have the choice of getting his pupils to carry out five completely different types of exercise, namely:

1. A detailed technical case study, involving physics-based calculations.
2. A detailed economic case study, again involving calculations.
3. A comparative case study involving no calculations.
4. A role-playing exercise simulating consumer research.
5. A highly competitive Power Station Game-type simulation game.

HYDROPOWER

Hydropower (p 134) is another multi-disciplinary multi-project pack, this time dealing with the theme of hydro-electric pumped storage (the system whereby electricity surplus to requirements during times of low demand is used to pump water from a low-level to a high-level reservoir, thus effectively 'storing' the electricity for later use during times of high demand). Like the Central Heating Game, it has its main basis in physics, but it can be used in the teaching of a wide range of other subjects (including geography, modern studies, science in society, engineering and economics) at both upper secondary and tertiary levels. The projects vary in length from one to three hours.

Hydropower had its origins in the Hydropower 77 inter-school competition run by the North of Scotland Hydro-Electric Board in 1976-77[67,68] (see Chapter 2). This was based on an extended multi-disciplinary design study in which teams of senior pupils had to examine six different potential sites for a 1000 MW hydro-electric pumped storage scheme, carry out technical and economic appraisals of the various possible schemes, and select the one they felt was the most promising, taking account of all relevant technical, economic, geographical, environmental and amenity factors.

While developing the scenario for Hydropower 77, the project authors realized that it would make an excellent source of in-depth case studies for use in schools and colleges at a variety of levels and in a variety of disciplines. The competition project was therefore converted into a multi-project pack containing six basic projects of different types. This was done by breaking the original project down into its constituent stages and preparing a hierarchy of resource materials (maps and data sheets) giving the results obtained at the end of each stage. These maps and data sheets, together with various items of introductory material, were then used as the basis of the individual projects, a given item sometimes serving as initial resource material and sometimes as debriefing material, depending on the nature of the project.

A list of the resource materials contained in the Hydropower package and a brief description of each of the six projects are given below, while the relationship between the individual projects and the hierarchy of resource materials is shown schematically in Figure 4.9.

List of resource materials

- — 12 copies of an introductory leaflet entitled 'Background Information about Hydro-Electric Pumped Storage'.
- — 3 copies each of four leaflets describing the North of Scotland Hydro-Electric Board's pumped storage schemes at Cruachan and Foyers and the Central Electricity Generating Board's schemes at Ffestiniog and Dinorwic in North Wales.
- — 6 copies each of three general maps of the area on which the scenario is based (showing general topography and communications, existing hydro-electric developments and land use).
- — 2 copies each of Site Maps 1a to 6a, giving detailed geographical information about each of the 6 possible pumped storage sites.
- — 6 copies each of Site Maps 1b to 6b, giving the same information as in Site Maps 1a to 6a plus detailed descriptions of the lay-outs of the schemes envisaged for the sites.
- — 6 copies of Data Sheet 1, giving the basic design parameters of the schemes envisaged for the 6 sites.
- — 6 copies of Data Sheet 2, giving detailed information about the dimensions of all underground workings for the various schemes.
- — 6 copies of Data Sheet 3, giving a detailed capital cost breakdown for each of the various schemes.
- — 6 copies of Data Sheet 4, giving a breakdown of present and likely future operating costs for the schemes.
- — 6 copies of Data Sheet 5, giving cumulative operating costs and an overall economic comparison of the various schemes.
- — 6 copies each of Project Sheets 1 to 6 (which describe the work of the 6 projects).
- — a teacher's guide, giving detailed instructions on how to run the various projects, plus suggestions as to how they could be used as a jumping-off point for further work.

Description of Project 1. This project is intended primarily for use with geography students, working either singly or (preferably) in small co-operative groups. It consists of a set of six design studies in which the geography of each of the six sites has to be examined with a view to determining the optimum lay-out and basic design parameters for a pumped storage scheme based on that site.

Description of Project 2. This project is designed for use with geography, modern studies or science in society students, again working either alone or in small co-operative groups. Each group is issued with a complete set of resource materials (except for Data Sheet 5), and has to use the information contained therein to carry out a comparative appraisal

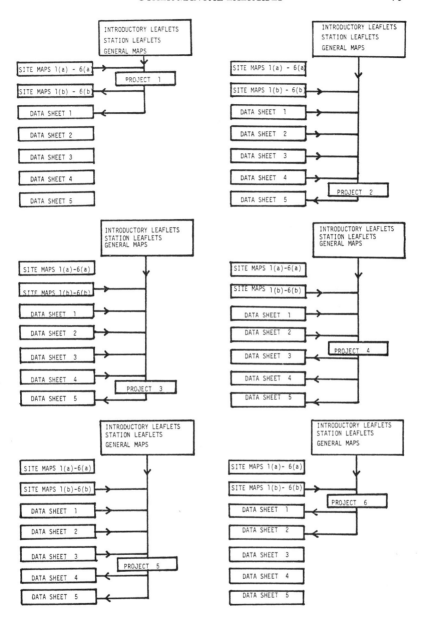

Figure 4.9: How the various Hydropower projects are related to the basic resource material in the pack

of the six possible schemes with a view to determining which is the most promising.

Description of Project 3. This project, which is intended for use with modern studies or science in society classes, is a role-playing exercise designed to highlight the conflict between technical and economic factors on the one hand and environmental and amenity considerations on the other that can arise in planning a major development such a pumped storage scheme. The class is divided into two competing groups, one of which has to pick the best scheme purely on technical and economic grounds, while the other has to decide which would be least harmful from an environmental and amenity point of view (the exercise has been designed in such a way that the two groups should always pick different schemes). A debate between the two groups then follows.

Description of Project 4. This project is primarily designed for use with economics students, working either singly or in small co-operative groups. Like Project 1, it consists of a set of six case studies in which the capital and operating costs of each of the six possible schemes have to be determined.

Description of Project 5. This project, which is again primarily designed for use with economics students, is a critical comparative study that involves determining the best scheme from a purely economic point of view. Like the previous project, the students can either work on their own or as small co-operative groups.

Description of Project 6. This, the longest project in the pack, consists of six extended design studies in which (a) the basic design parameters and (b) the dimensions of the underground workings have to be determined for the six possible schemes. Each case study is divided into two sections in such a way that a class can either tackle both sections end on or limit work to one or the other. The project is designed for use with physics or engineering students, again working either singly or in small co-operative groups.

Hydropower is a good example of the way in which a highly complex and demanding multi-disciplinary case study can, by use of the multi-project structure, be broken down into smaller projects suitable for use in the teaching of the various academic subjects on which the case study is based. Like the Central Heating Game, it also shows how a wide range of different projects can be based on a common set of resource materials.

Chapter 5

Computer-based exercises

Apart from its obvious role as a 'supercalculator', there are two basic ways in which the computer can contribute to the teaching/learning process, namely as a *tutorial teaching machine* and as a *laboratory-substitute.*

In the 'tutor' mode, the student interacts with the computer, which is programmed to react to responses to questions that it sets. This style of learning is directly descended from the programmed instruction movement of the 1950s and 1960s. It is essentially similar to branching programmed learning, but is capable of being much more sophisticated than the latter because of the greater flexibility and data handling capacity of computers compared with teaching machines or programmed texts.

In the 'laboratory' mode, the computer is essentially a learning resource rather than a direct teaching device. Basically, it simulates a laboratory situation, being used to model experiments, provide data bases, set problem-solving exercises, and so on. Using the computer, students can, for example, find out what happens to physical, chemical, biological or industrial systems under varying conditions that they themselves specify, thus providing much greater flexibility than would ever be practicable in a conventional laboratory.

The computer can, of course, also be used in a combination of these two basic roles, and many of the most useful computer-assisted learning exercises do in fact incorporate both forms of usage. The five 'laboratory' simulations described in this chapter, for example, all have a certain amount of 'tutorial' element in their structure.

In Britain, a five-year government-funded National Development Programme in Computer Assisted Learning (NDPCAL) was completed in 1977 at a cost of around £2.5 million.[84] The programme consisted of over 30 separate projects, covering a wide range of disciplines. Many of the projects involved several colleges and universities working co-operatively. The science-based projects generated material in both the tutor and the laboratory modes.

In this chapter, we will consider some of the laboratory mode exercises, since these have the closest links with the gaming/simulation/case study field. To date, their main use has been in the development of higher cognitive skills, including problem-solving and decision-making skills. Another reason for concentrating on the laboratory side of computer-assisted learning is that this may well be the area in which much of the most important future development takes place. The role of the computer in education has yet to be firmly established, and it seems doubtful that computer-assisted learning will ever flourish if it is seen merely as an expensive alternative to the classroom teacher. If it is to become an established weapon in our educational armoury, it is vital that we identify and investigate uses to which the computer is uniquely suited. The computerized laboratory simulation certainly seems to be one such area.

In the past, two important restrictions on the widespread use of computers in education have been the problems of non-availability of hardware and lack of transferability of software. A further constraint on the use of some computer simulations has been the necessity to have a video display unit available. The various hardware problems are now being greatly reduced by the advent of cheap, portable and easy-to-use microcomputers, which will no doubt encourage much more widespread adoption of computers into the curricula of schools, colleges and universities. The software problem is also becoming progressively less serious since many of the software packages that have been produced in recent years are generally available, and the designers are usually only too willing to give advice regarding their use and applicability.

As mentioned in Chapter 2, areas in which computer simulations could make a particularly valuable contribution to science education include the following:

— situations where a conventional experiment is either extremely difficult or impossible;
— situations where experimental apparatus is either not readily available or too complicated or expensive for general laboratory use;
— situations where actual experimental work could be dangerous, or would cause unnecessary suffering;
— situations where a conventional experiment would take an unacceptably long time to complete.

One further way in which the computer can be used in science-based simulation exercises is in the role of the 'manager' of the exercise. Here, the computer *directs* the sequence of work and carries out most of the heavy calculations, leaving the students free to make the various decisions that the exercise involves in the course of conventional group discussions.

Specific examples of the application of computers to each of the above areas will now be examined, although it should be realized that there can, in practice, be considerable overlap between the various categories listed.

Situations where a conventional experiment is either extremely difficult or impossible (Satellite Motion)

An example of this type of computer simulation is Satellite Motion (p 142) — a physics-based exercise produced as part of the Computers in the Undergraduate Science Curriculum (CUSC) Project in the National Development Programme. The program simulates the motion of a satellite launched from the earth's surface.

In the first part of the exercise, the student selects the angular momentum of the satellite, and the resultant plots of gravitational

potential energy and orbital kinetic energy as functions of radial position
are displayed on the screen of a graphics terminal. The computer then
calculates and displays the corresponding orbit. Figure 5.1 shows the
energy levels and the resulting elliptical orbits for a specified angular
momentum of the satellite. This part of the program is intended to
increase the student's understanding of the relationship between the
angular momentum and energy of a satellite and the shape of the resulting
orbit.

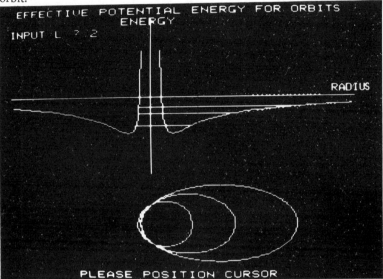

*Figure 5.1: The energy diagram for a satellite with an angular momentum of
two 'units' showing the different energy levels corresponding to three elliptical
orbits (Satellite Motion).*

The second part of the exercise simulates the trajectory of a satellite
launched from the earth, and allows the user to change the velocity of the
satellite after half an orbit has been traced out. The student is encouraged
to try a variety of different manoeuvres, such as transferring the satellite
to a higher or lower circular orbit (Hohmann transfer), or making it
escape from the earth's gravitational field, as shown in Figure 5.2

It is also possible to extend the scope of the exercise to include, for
example, the effect of the solar wind on the satellite's motion, or to
consider the 'three-body problem' that results from taking account of the
moon's gravitational field as well as that of the earth.

This package simulates an experiment which it would never be
practicable to perform in real-life. It is an excellent example of the way
in which the computer can be used to model inaccessible physical
situations, thus allowing students to gain an insight which would be
hard to achieve using more conventional methods such as manual
calculations.

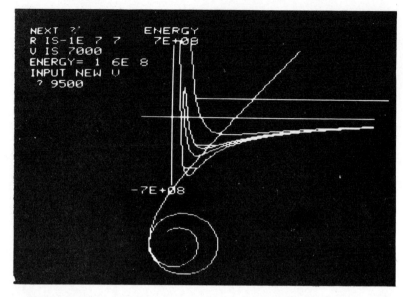

Figure 5.2: Energy diagrams showing the successive orbits of a satellite that eventually escapes from the earth's gravitational field (Satellite Motion).

Situations where experimental apparatus is either not readily available or too complicated or expensive (HABER)

Many scientific experiments are not carried out in school or college laboratories because of the complexity of equipment required or the intrinsic difficulty of achieving the desired experimental conditions. Examples include experiments in nuclear and particle physics and many industrial processes.

For example, the *Haber Process* (an industrial chemical process used to convert nitrogen and hydrogen into ammonia) is extremely difficult to carry out on a laboratory scale due to the high pressures involved. HABER (p 161) is a computer simulation which uses a mathematical model of the process. It is designed to help students discover the effects of altering various conditions (temperature, pressure, catalyst and reactant concentration ratios) on the course of the reaction. The package, which was produced as part of the Chelsea Science Simulation Project, is designed for use at senior secondary and lower tertiary levels. It comes in the form of A4 teacher and student booklets. The two-part computer program is in BASIC, and is designed to run on most commonly used computer systems.

Investigation 1 looks at the effect of varying temperature, pressure and the initial molar ratio of hydrogen to nitrogen on the percentage yield of ammonia (ie *thermodynamic* factors). The student booklet describes the prior knowledge that is required in order to carry out the exercise effectively. When interacting with the computer, the students enter simple replies to the computer's questions. Figure 5.3 shows a sample print-out, with the user's inputs underlined.

Students follow up this first part of the exercise by carrying out a detailed study of the Haber industrial process, with the information gained from the program being used to answer further questions set in the student booklet.

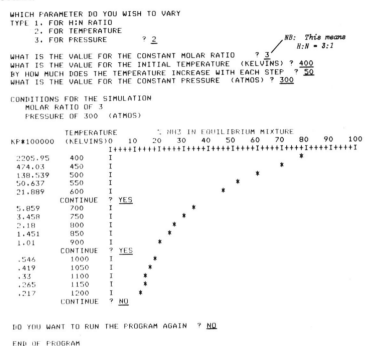

```
WHICH PARAMETER DO YOU WISH TO VARY
TYPE 1. FOR H:N RATIO
     2. FOR TEMPERATURE
     3. FOR PRESSURE          ? 2                    NB: This means
                                                         H:N = 3:1
WHAT IS THE VALUE FOR THE CONSTANT MOLAR RATIO    ? 3
WHAT IS THE VALUE FOR THE INITIAL TEMPERATURE  (KELVINS) ? 400
BY HOW MUCH DOES THE TEMPERATURE INCREASE WITH EACH STEP  ? 50
WHAT IS THE VALUE FOR THE CONSTANT PRESSURE  (ATMOS) ? 300

CONDITIONS FOR THE SIMULATION
     MOLAR RATIO OF 3
     PRESSURE OF 300  (ATMOS)

                TEMPERATURE     % NH3 IN EQUILIBRIUM MIXTURE
KP*100000      (KELVINS)0    10    20    30    40    50    60    70    80    90    100
                        I++++I++++I++++I++++I++++I++++I++++I++++I++++I++++I
2205.95          400    I                                                *
 474.03          450    I                                            *
 138.539         500    I                                       *
  50.637         550    I                                   *
  21.889         600    I                              *
              CONTINUE  ? YES
   5.859         700    I                    *
   3.458         750    I                  *
   2.18          800    I               *
   1.451         850    I              *
   1.01          900    I          *
              CONTINUE  ? YES
    .546        1000    I         *
    .419        1050    I         *
    .33         1100    I       *
    .265        1150    I       *
    .217        1200    I      *
              CONTINUE  ? NO

DO YOU WANT TO RUN THE PROGRAM AGAIN  ? NO

END OF PROGRAM
```

Figure 5.3: Sample print-out from Investigation 1 of HABER program

Investigation 2 looks at the effect of varying temperature, pressure and catalyst on the rate of attaining chemical equilibrium (ie *kinetic* factors).

Again the basic pre-knowledge required is detailed in the student booklet. Figure 5.4 shows a sample print-out from this part of the program, the user's responses to the questions again being underlined.

The student concludes the exercise by answering further questions listed in the student booklet.

Situations where actual experimental work could be dangerous, or would cause suffering (DYE)

If there is a serious risk of accident or health hazard associated with an experiment (when working with explosive, highly radioactive or extremely toxic substances, for example), or when the subject of the experiment (human or animal) might undergo undue suffering, a computer simulation has obvious advantages.

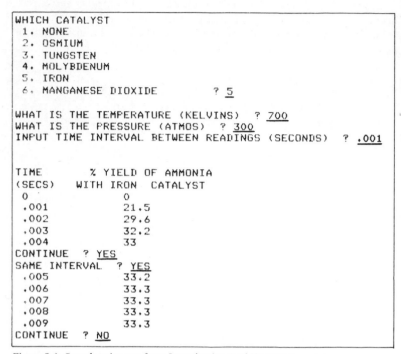

```
WHICH CATALYST
  1. NONE
  2. OSMIUM
  3. TUNGSTEN
  4. MOLYBDENUM
  5. IRON
  6. MANGANESE DIOXIDE            ? 5

WHAT IS THE TEMPERATURE (KELVINS)   ? 700
WHAT IS THE PRESSURE (ATMOS)   ? 300
INPUT TIME INTERVAL BETWEEN READINGS (SECONDS)   ? .001


TIME        % YIELD OF AMMONIA
(SECS)     WITH IRON  CATALYST
 0             0
 .001         21.5
 .002         29.6
 .003         32.2
 .004         33
CONTINUE   ? YES
SAME INTERVAL   ? YES
 .005         33.2
 .006         33.3
 .007         33.3
 .008         33.3
 .009         33.3
CONTINUE   ? NO
```

Figure 5.4: Sample print-out from Investigation 2 of HABER program

An example of this type of simulation is the biology-based program DYE (p 182) from the Computers in the Undergraduate Science Curriculum Project. This is used to teach the principles involved in the indicator dilution method for measuring cardiac output.

The use of a computer simulation for presentation of this topic completely eliminates the danger to human or animal life that would be associated with an actual experiment. The user selects the amount of indicator dye to be injected into a vein, and also how frequently blood samples are to be collected from an artery. The program displays the dye concentration in these samples as a function of time, and a moveable cursor can then be used to determine the area under the graph and hence the cardiac output. Figure 5.5 shows a typical display on the graphics terminal. In this case, the user decided to sample every second. Using the moveable cursor, the co-ordinates of some of the points have been printed out, along with the logarithm of the y co-ordinates.

It is also possible to use the package to become familiar with the standard technique for correcting the dye recirculation, by plotting values of the logarithm of the concentration against time. First, the program sets the cardiac output at a random value between pre-defined limits. The user is then invited to input an estimate of cardiac output,

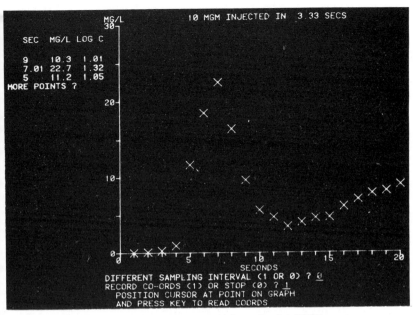

Figure 5.5: Graphics display for cardiac output determination (DYE)

and, if this is within 5 per cent of the set value, he may proceed to the second part of the program. This allows variation (within limits) of some of the factors which alter the form of the dye dilution curve (see Figure 5.6). These factors include such things as the size of the heart, the volume of the circulation, and the rate of dye injection. The dye dilution curve for particular values of cardiac output specified by the user can also be displayed.

A further advantage of this particular simulation is that the user has far greater control over the variables involved than would ever be the case in an actual experiment.

Situations where a conventional experiment would take an unacceptably long time to complete (OPERON)

In experiments with large time scales (such as investigations into population dynamics or work with long-lived radioactive materials), the time span can be reduced to manageable proportions by using a computer simulation.

An example from the Computers in the Undergraduate Science Curriculum Project is the biology-based simulation OPERON (p 186) which looks at the induction of gene activity. In this exercise, the Jacob-Monod model of genetic induction is introduced, and is developed by means of three options available to the user:

1. Using a single (pre-defined) bacterial strain. Here, the user selects (arbitrarily) one strain from 12 available, and assays this for

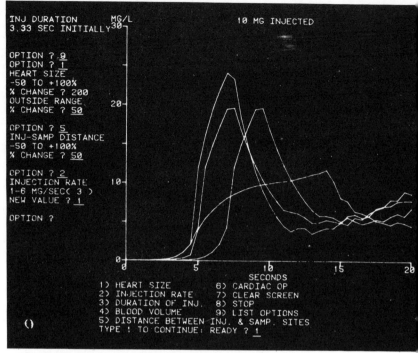

Figure 5.6: Graphics display for dye dilution curve (DYE)

enzymic activity either with or without the inducer. Sampling time, total assay time, and time of introduction of the inducer can be varied by the user.

2. Two strains of bacteria are selected, one as donor and the second as recipient. A 'genetic cross' is made, to construct a heterogenote, and enzymic activity is again assayed.

3. Users are invited to create their own bacterial strain, by specifying both the order of the genes within the operon and the content of the genes. Figure 5.7 shows a teletypewriter output for this part of the program. The user defines the gene order and the content of the sites in a bacterial strain, then assays for enzyme activity.

When choosing the desired option, there is an opportunity to type a 'Help' code which prints out a partial characteristic of each of the 12 available strains and gives hints on how to choose the donor and recipient strains. There are seven classes of results (with two strains), and there is an option to allow the 'assay' of cross-reacting material produced by mutant structural genes.

The OPERON package allows the user to combine genetic elements at will. Such an option is not usually available to students because of technical and time constraints.

```
IF YOU REQUIRE HELP THEN TYPE 999 AFTER THE REQUEST

  YOU CAN
      1) ANALYSE A SINGLE DEFINED STRAIN
      2) CROSS TOGETHER TWO DEFINED STRAINS AND ANALYSE
              THE RESULTING HETEROGENOTE
  OR
      3) CREATE YOUR OWN HETEROGENOTE AND ANALYSE THAT

WHICH OPTION  (1,2 OR 3) ?  3

  KEEP A RECORD OF YOUR INPUTS
  ----------------------------

      GENE ORDER
      ----------

  1)   Z--O--I      2)  I--O--Z   Z=STRUCTURAL SITE
  2)   O--Z--I      4)  I--Z--O   O=OPERATOR SITE
  5)   O--I--Z      5)  Z--I--O   I=REGULATORY SITE

  WHICH ORDER   A)   DONOR ?  2  B)   RECIPIENT ?  1

  CONTENTS OF THE SITES
  ---------------------
  TYPE INTEGERS
  ABSENT=0 : FUNCTIONAL=1
  MUTANT (NON FUNCTIONAL)=2
  MUTANT (SUPER-REPRESSOR)=3 *** REGULATORY SITE ONLY

  STRUCTURAL SITE
   A) DONOR ?  1   B) RECIPIENT ?  1
  OPERATOR SITE
   A) DONOR ?  1   B) RECIPIENT ?  1
  REGULATORY SITE
   A) DONOR ?  3   B) RECIPIENT ?  1

TOTAL TIME OF ENZYME ASSAY
(1--->180 MINS) ?  120
MINS BETWEEN SAMPLES
(1--->60 MINS) ?  12
DO YOU WANT THE INDUCER
(TYPE 1 FOR YES 2 FOR NO) ?  1

AT WHAT TIME DO YOU WANT THE INDUCER
(FROM 1 TO 180 MINS) ?  15

      TIME         RATE OF ENZYME
      (MINS)        PRODUCTION
       0             0
       12            0
       24            3.6
       36            8.4
       48            10
       60            7.9
       72            4.5
       84            2.0
       96            1.5
       108           .86
       120           .49
INDUCER ADDED AFTER 15 MINS

WHICH OPTION  (1,2 OR 3) ?
```

Figure 5.7: Teletypewriter output from OPERON

Computer-managed simulations (The Alkali Industry)

Although by far the most common function that computers fulfil in the
simulation field is that of an interactive teaching resource (as in the
examples described so far), they have also occasionally been used in a
managerial role. Here, the computer is used to *direct* the sequence of
work (and carry out most of the heavy calculations) in what was originally
a manual simulation or case study.[85]

One problem often experienced by designers and users of industrial
simulations is that such exercises generally have to be greatly simplified
in order to enable them to fit into the relatively short periods of time
available for their completion. This simplification can, in extreme cases,
rob the simulation of much of its realism and vitality. In order to allow
students more time for group discussion and decision making, computerized
versions of a number of exercises of this type have now been produced. In
these, the problem of over-simplification need not arise, since the
computer can take over the main calculation burden.

An example of an exercise that has been converted in this way is
The Alkali Industry (p 173) which was originally developed in the
University of Glasgow as a manual simulation. The exercise is based on
the exploitation of a hypothetical find of salt deposits, supposedly located
in the Scottish county of Dumfries.

The amount of calculation originally needed to form the basis of the
decision-making process tended to obscure the main purpose of the
exercise, even at first-year undergraduate level. By adapting the exercise
to the computer, however, it has been possible not only to overcome this
inherent difficulty, but also to scale the simulation down, while still
retaining its essential realism. This has allowed it to be used successfully
with secondary school pupils aged 15 and over.

The computerized simulation has more or less the same structure as the
original, and also has the same basic educational aims. The program is
written in Extended BASIC and consists of six sub-programs. The entire
exercise takes around three hours to complete.

During the course of the simulation, the students make a series of
decisions based on the assumption that the salt will be used to produce
sodium carbonate and related compounds. First, they choose a site for the
proposed plant, then decide (on the basis of known demands for sodium
carbonate, sodium hydroxide and chlorine extrapolated to the year 2000)
when construction work should start.

These tasks were originally done manually but, in the computerized
version, they are completed using sub-program 1. This predicts demands
for the three commodities for any year up to 2000, and also calculates
the minimum capacity that the students' plant should have in that year.

For the purpose of the simulation, it is assumed that there are two
possible processes whereby the required products may be manufactured,
namely the *Solvay* and *Castner-Kellner* processes. Sub-program 2 enables
the capital costs of plants of different capacity to be calculated for the

two processes, both for new 'greenfield' sites and for sites that have already been developed.

Having costed the proposed plant, material costs are now calculated. This part of the exercise is not computerized, since it was felt that the students would gain more by working out these costs for themselves. At the end of this section, the students are expected to have calculated the total profit (using both the Solvay and the Castner-Kellner processes) on the sale of one tonne of sodium carbonate and its marketable by-products.

Next, the students have to decide on the best method of production by drawing up cash-flow diagrams. Data are fed into the computer, whose output comes in the form of two columns headed 'Year' and 'Total Profit'. This information is then converted into a cash-flow graph by the students.

Sub-program 4 allows the option of a pilot plant being built before the major plant, thus enabling the effect of this on the cash-flow diagram to be investigated.

Sub-programs 5 and 6 describe two refinements which have been added to the original content of the exercise. Having produced an optimum cash-flow diagram, students can use sub-program 5 to cost the same plant on the basis of the figures for 1969. They can therefore investigate the financial consequences of making the same decisions in that year. This enables the students to see that a combination of inflation and rapidly rising energy costs have resulted in a relative shift in profitability from the Castner-Kellner process to the Solvay process over the intervening period.

One of the major alkali production processes now used in Britain is the *diaphragm cell* process which overcomes the mercury pollution problems associated with the Castner-Kellner process. Sub-program 6 allows students to work out the cost of a diaphragm cell plant of known capacity, and also to calculate the associated material costs.

The exercise concludes with a general discussion section, covering the effects of the students' decisions on the environment, employment, and profit. A balance has then to be found between these conflicting factors.

A comparative evaluation has shown that students seem to prefer the computerized version of The Alkali Industry to the original manual version. No serious difficulties have been encountered in connection with the use of teletype computer terminals, and the exercise has now been successfully used with a wide range of students — from chemistry undergraduates down to 15-year-old schoolchildren. It could serve as a useful paradigm for similar conversion of other exercises.

Selecting, adapting and using existing exercises

Having established why science-based games, simulations and case studies are educationally useful and examined the different types of exercise, we shall now turn to a number of more practical topics. This chapter will deal with the range of materials currently generally available, suggesting how exercises can be selected for specific educational purposes and, if not exactly suitable in their original form, adapted to meet the exact requirements of the user. It will then offer advice on how such exercises can be used to best effect in teaching. Chapter 7 will show how teachers and lecturers can design their own games, simulations and case studies if what they want is not already available. Finally, Chapter 8 will discuss the evaluation of gaming and simulation materials.

Selecting exercises for specific purposes

When selecting a science-based game, simulation or case study for a particular educational purpose, it is necessary to ask oneself a number of basic questions. Many of these may seem obvious, especially to readers who have some familiarity with the medium, but it is felt necessary to spell them out in detail because of the importance of correct selection in ensuring that intended educational outcomes are in fact properly achieved.

The selection process can be conveniently divided into three stages, the first of which is the determination of the broad purpose for which the exercise is to be used. This can be done using the algorithm given in Figure 6.1, which is, we hope, self-explanatory.

The second stage of the selection process is the identification of exercises that might be suitable for the purpose in mind. This can be done by using the lists (in Figures 6.2 to 6.6) to which the potential user is directed by the algorithm or, in some cases, by carrying out a systematic search through the appropriate section of Part 2.

The third and final stage involves examining the various possible exercises identified in the second stage in order to determine which (if any) is best suited to fulfil the specific role the potential user has in mind. This can be done by referring to the data sheets contained in Part 2, which provide detailed information about all commercially (or otherwise generally) available science-based exercises known to the authors at the time of writing.

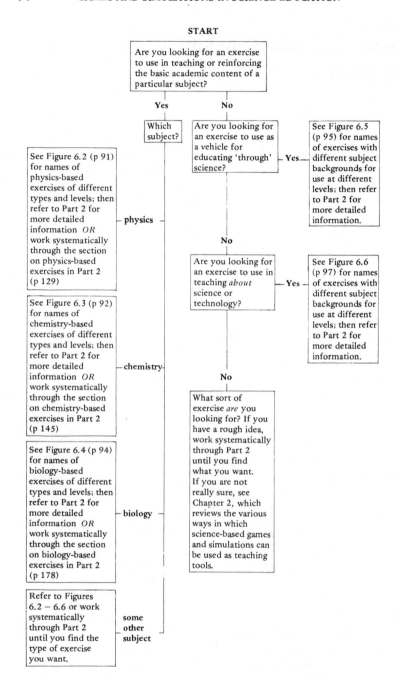

Figure 6.1: Guide to the selection of science-based exercises for specific educational purposes

Name of exercise	Format	12	13	14	15	16	17	18	19	20	21	22	Page
Safety Snakes and Ladders	Board	━	━	━									197
Lab Apparatus	Card	━	━	━	━								195
Circuitron	Board	━	━	━	━	━	━	━					131
389/Science Concepts	Card	━	━	━	━								198
Energy for the Future	Manual			━	━								158
Isotopes	Manual			━	━								162
Energy — Past, Present and Future	Manual			━	━	━							158
Science Sense	Board			━	━	━							199
Gravitational Fields	Computer			━	━	━							134
The Electric Circuit	Board			━	━	━							143
Energy Conversions	Board			━	━	━							132
The Energy Problem	Manual			━	━	━							174
Physics Crosswords	Manual			━	━	━							138
Planetary Motion	Computer			━	━	━							139
Ekofisk — One of a Kind	Manual			━	━	━							193
Alternative Energy Project	Manual				━	━	━						129
Power Station Project	Manual				━	━	━						140
Central Heating Project	Manual				━	━	━						131
The Solar System	Board				━	━	━						144
Keeping Warm	Board				━	━	━						136
Radioactive Decay	Computer				━	━	━						141
Isotopes in Our Lives	Manual					━	━	━					162
Photoelectric Effect	Computer					━	━	━					138
Capacitor Discharge	Computer					━	━	━					129
Gaseous Diffusion	Computer					━	━	━					133
Central Heating Game	Manual					━	━	━					130
Power for Elaskay	Manual					━	━	━					139
The Power Station Game	Manual					━	━	━					144
SCATTER	Computer					━	━	━	━				142
INTERP	Computer					━	━	━	━				135
NEWTON	Computer					━	━	━	━				137
Hydropower	Manual					━	━	━	━				134
Mass Spectrometer	Computer						━	━	━	━	━		136
Free Fall with Air Resistance	Computer							━	━	━			133
Satellite Motion	Computer							━	━	━			142
Flow Through a Nozzle	Computer							━	━	━			133
Particle in a Potential Well	Computer								━	━	━	━	166
RANDOM	Computer								━	━	━	━	170
PBARR	Computer									━	━	━	137
PSTEP	Computer									━	━	━	141
PWELL	Computer									━	━	━	141

Figure 6.2: Exercises suitable for use in teaching physics

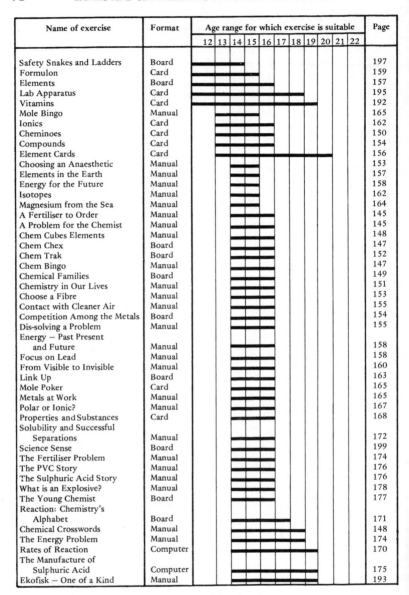

Name of exercise	Format	Age range for which exercise is suitable	Page
Safety Snakes and Ladders	Board		197
Formulon	Card		159
Elements	Board		157
Lab Apparatus	Card		195
Vitamins	Card		192
Mole Bingo	Manual		165
Ionics	Card		162
Cheminoes	Card		150
Compounds	Card		154
Element Cards	Card		156
Choosing an Anaesthetic	Manual		153
Elements in the Earth	Manual		157
Energy for the Future	Manual		158
Isotopes	Manual		162
Magnesium from the Sea	Manual		164
A Fertiliser to Order	Manual		145
A Problem for the Chemist	Manual		145
Chem Cubes Elements	Manual		148
Chem Chex	Board		147
Chem Trak	Board		152
Chem Bingo	Manual		147
Chemical Families	Board		149
Chemistry in Our Lives	Manual		151
Choose a Fibre	Manual		153
Contact with Cleaner Air	Manual		155
Competition Among the Metals	Board		154
Dis-solving a Problem	Manual		155
Energy — Past Present and Future	Manual		158
Focus on Lead	Manual		158
From Visible to Invisible	Manual		160
Link Up	Board		163
Mole Poker	Card		165
Metals at Work	Manual		165
Polar or Ionic?	Manual		167
Properties and Substances	Card		168
Solubility and Successful Separations	Manual		172
Science Sense	Board		199
The Fertiliser Problem	Manual		174
The PVC Story	Manual		176
The Sulphuric Acid Story	Manual		176
What is an Explosive?	Manual		178
The Young Chemist	Board		177
Reaction: Chemistry's Alphabet	Board		171
Chemical Crosswords	Manual		148
The Energy Problem	Manual		174
Rates of Reaction	Computer		170
The Manufacture of Sulphuric Acid	Computer		175
Ekofisk — One of a Kind	Manual		193

Figure 6.3: Exercises suitable for use in teaching chemistry

Name of Exercise	Format	Age range for which exercise is suitable											Page
		12	13	14	15	16	17	18	19	20	21	22	
The Protein Problem	Manual				■	■	■						175
Take Your Choice	Manual				■	■	■						172
Case Studies in Chemical Engineering	Manual				■	■	■						146
Chemistry Case Studies	Manual				■	■	■						150
Isotopes in Our Lives	Manual					■	■						162
Public Inquiry Project	Manual					■	■	■					169
Chemical Element Game	Computer					■	■	■	■				149
Dental Health Project	Manual					■	■	■					193
Electrochemical Cells	Computer					■	■	■	■				156
Homogeneous Equilibrium	Computer					■	■	■					161
Gas Chromatography	Computer					■	■	■					160
Lattice Energy	Computer					■	■	■					163
HABER	Computer					■	■	■					161
RKINET	Computer					■	■	■					171
Point Fields	Manual					■	■	■					166
The Alkali Industry	Computer					■	■	■	■	■	■		173
Fluoridation?	Manual				■	■	■	■	■	■			194
Proteins as Human Food	Manual						■	■					168
Chemsyn	Card					■	■	■					151
What Happens When The Gas Runs Out?	Manual						■	■					177
The Amsyn Problem	Manual						■	■	■				173
Particle in a Potential Well	Computer							■	■	■	■	■	166
RANDOM	Computer							■	■	■	■	■	170
Batch or Flow	Manual								■	■	■	■	146
PBARR	Computer								■	■	■	■	137
PSTEP	Computer								■	■	■	■	141
PWELL	Computer								■	■	■	■	141
Polywater	Manual									■	■	■	167

Figure 6.3 (continued)

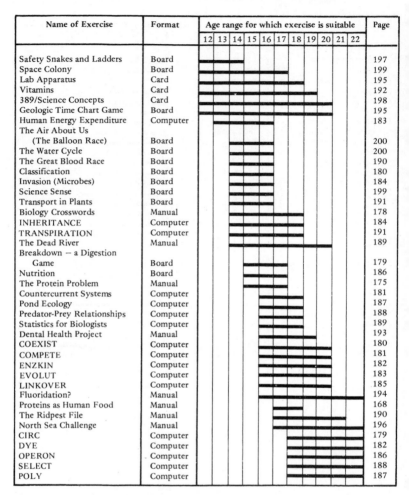

Figure 6.4: Exercises suitable for use in teaching biology

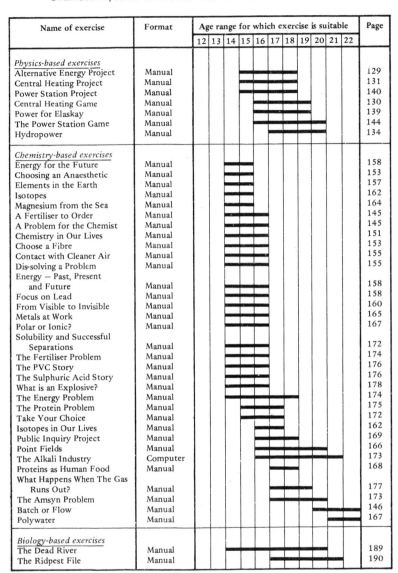

Name of exercise	Format	12	13	14	15	16	17	18	19	20	21	22	Page
Physics-based exercises													
Alternative Energy Project	Manual				■	■	■	■					129
Central Heating Project	Manual				■	■	■	■					131
Power Station Project	Manual				■	■	■	■					140
Central Heating Game	Manual					■	■	■	■				130
Power for Elaskay	Manual					■	■	■	■				139
The Power Station Game	Manual					■	■	■	■				144
Hydropower	Manual						■	■	■	■			134
Chemistry-based exercises													
Energy for the Future	Manual				■	■							158
Choosing an Anaesthetic	Manual				■	■							153
Elements in the Earth	Manual				■	■							157
Isotopes	Manual				■	■							162
Magnesium from the Sea	Manual			■	■	■							164
A Fertiliser to Order	Manual				■	■	■						145
A Problem for the Chemist	Manual				■	■	■						145
Chemistry in Our Lives	Manual				■	■	■						151
Choose a Fibre	Manual				■	■	■						153
Contact with Cleaner Air	Manual				■	■	■						155
Dis-solving a Problem	Manual				■	■	■						155
Energy — Past, Present and Future	Manual				■	■	■						158
Focus on Lead	Manual				■	■	■						158
From Visible to Invisible	Manual				■	■	■						160
Metals at Work	Manual				■	■	■						165
Polar or Ionic?	Manual				■	■	■						167
Solubility and Successful Separations	Manual				■	■	■						172
The Fertiliser Problem	Manual				■	■	■						174
The PVC Story	Manual				■	■	■						176
The Sulphuric Acid Story	Manual				■	■	■						176
What is an Explosive?	Manual				■	■	■						178
The Energy Problem	Manual					■	■	■					174
The Protein Problem	Manual					■	■	■					175
Take Your Choice	Manual				■	■	■	■					172
Isotopes in Our Lives	Manual					■	■	■					162
Public Inquiry Project	Manual						■	■	■				169
Point Fields	Manual						■	■	■				166
The Alkali Industry	Computer						■	■	■	■			173
Proteins as Human Food	Manual							■	■				168
What Happens When The Gas Runs Out?	Manual							■	■	■			177
The Amsyn Problem	Manual							■	■	■			173
Batch or Flow	Manual								■	■	■		146
Polywater	Manual								■	■	■		167
Biology-based exercises													
The Dead River	Manual			■	■	■	■	■	■	■			189
The Ridpest File	Manual					■	■	■	■	■			190

Figure 6.5: Exercises suitable for education 'through' science

Name of Exercise	Format	Age range for which exercise is suitable											Page
		12	13	14	15	16	17	18	19	20	21	22	
Other exercises													
Space Colony	Board		■	■	■	■	■	■	■				199
389/Science Concepts	Card	■	■	■	■	■	■						198
Offshore Oil Board Game	Board			■	■	■	■	■	■				196
Dental Health Project	Manual						■	■	■	■			193
Fluoridation?	Manual					■	■	■	■	■	■	■	194
North Sea Challenge	Manual						■	■	■	■	■	■	196

Figure 6.5 (continued)

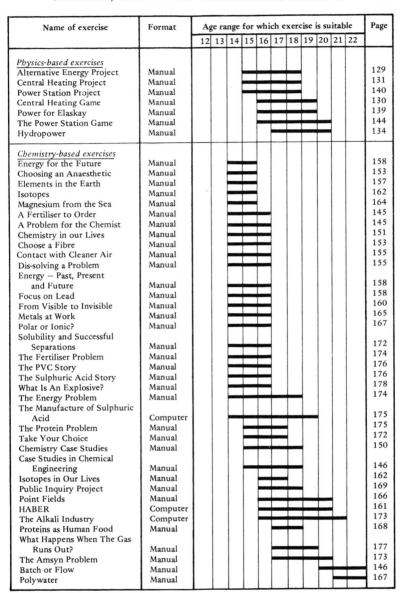

Name of exercise	Format	12	13	14	15	16	17	18	19	20	21	22	Page
Physics-based exercises													
Alternative Energy Project	Manual				██	██	██						129
Central Heating Project	Manual				██	██	██						131
Power Station Project	Manual			██	██	██							140
Central Heating Game	Manual					██	██	██					130
Power for Elaskay	Manual					██	██	██					139
The Power Station Game	Manual					██	██	██					144
Hydropower	Manual						██	██					134
Chemistry-based exercises													
Energy for the Future	Manual			██									158
Choosing an Anaesthetic	Manual			██									153
Elements in the Earth	Manual			██									157
Isotopes	Manual			██									162
Magnesium from the Sea	Manual			██									164
A Fertiliser to Order	Manual			██	██								145
A Problem for the Chemist	Manual			██	██								145
Chemistry in our Lives	Manual			██	██								151
Choose a Fibre	Manual			██	██								153
Contact with Cleaner Air	Manual			██	██								155
Dis-solving a Problem	Manual			██	██								155
Energy — Past, Present and Future	Manual			██	██								158
Focus on Lead	Manual			██	██								158
From Visible to Invisible	Manual			██	██								160
Metals at Work	Manual			██	██								165
Polar or Ionic?	Manual			██	██								167
Solubility and Successful Separations	Manual			██	██								172
The Fertiliser Problem	Manual			██	██								174
The PVC Story	Manual			██	██								176
The Sulphuric Acid Story	Manual			██	██								176
What Is An Explosive?	Manual			██	██								178
The Energy Problem	Manual			██	██								174
The Manufacture of Sulphuric Acid	Computer			██	██	██	██						175
The Protein Problem	Manual				██	██							175
Take Your Choice	Manual				██	██							172
Chemistry Case Studies	Manual				██	██	██						150
Case Studies in Chemical Engineering	Manual				██	██	██						146
Isotopes in Our Lives	Manual					██	██						162
Public Inquiry Project	Manual					██	██						169
Point Fields	Manual					██	██	██					166
HABER	Computer					██	██	██					161
The Alkali Industry	Computer					██	██	██	██				173
Proteins as Human Food	Manual					██							168
What Happens When The Gas Runs Out?	Manual					██	██						177
The Amsyn Problem	Manual					██	██						173
Batch or Flow	Manual								██	██			146
Polywater	Manual								██	██			167

Figure 6.6: Exercises for teaching about science and technology

Name of exercise	Format	Age range for which exercise is suitable											Page
		12	13	14	15	16	17	18	19	20	21	22	
Biology-based exercises													
Classification	Board			▬	▬	▬	▬						180
The Dead River	Manual			▬	▬	▬							189
The Ridpest File	Manual						▬	▬	▬	▬			190
Other exercises													
Space Colony	Board	▬	▬	▬	▬	▬							199
Lab Apparatus	Card	▬	▬	▬	▬	▬	▬						195
Ekofisk — One of a Kind	Manual	▬	▬	▬	▬								193
389/Science Concepts	Card	▬	▬	▬	▬								198
Geologic Time Chart Game	Board	▬	▬	▬	▬								195
Offshore Oil Board Game	Board			▬	▬	▬							196
Dental Health Project	Manual						▬	▬	▬	▬			193
Queries 'n' Theories	Board						▬	▬	▬	▬			197
Fluoridation?	Manual						▬	▬	▬				194
North Sea Challenge	Manual						▬	▬	▬	▬	▬		196

Figure 6.6 (continued)

The selection process can best be illustrated by looking at specific examples.

EXAMPLE 1

A teacher of chemistry in a secondary school wishes to find an exercise suitable for use in reinforcing basic work on the formation of ionic and covalent compounds with a class of 12 to 13 year olds. The exercise must be inexpensive (because his funds are strictly limited), relatively short (so that it can be easily fitted into the curriculum), and fairly simple (because his pupils are not of the highest academic calibre). Also, it must be manual rather than computer-based, since he does not have multiple access to computer facilities.

Stage 1. On referring to Figure 6.1 (the basic algorithm), the teacher is immediately directed to Figure 6.3 by answering 'yes' and 'chemistry' to the first two questions.

Stage 2. On referring to Figure 6.3, the teacher finds a list of all currently available exercises suitable for use in teaching chemistry that shows, for each exercise, (a) the format (card game, board game, manual exercise or computer-based exercise), (b) the range of ages over which it can be used, and (c) the page reference in Part 2. From this, he draws up the following short list of exercises of the appropriate level and format whose names indicate that they *might* be the sort of thing he is looking for:

Formulon (p 159)	Cheminoes (p 150)
Elements (p 157)	Compounds (p 154)
Ionics (p 162)	Element Cards (p 156)

Stage 3. The teacher now refers to the data sheets on these exercises and reaches the following conclusions about each.

Formulon — a simple, inexpensive card game dealing with the formation of ionic and covalent compounds. It therefore appears to be ideally suitable.

Elements — a simple board game designed to reinforce class work on the periodic table. It does not deal with the formation of compounds, however, and is therefore unsuitable.

Ionics — a card game dealing with *ionic* formulae and equations only. It is therefore unsuitable.

Cheminoes — a simple, inexpensive dominoes-type game dealing with the formation of ionic and covalent compounds. It therefore appears to be ideally suitable for the purpose the teacher has in mind.

Compounds — a simple game designed to reinforce work on the formation of basic chemical compounds. It therefore appears to be ideally suitable.

Element Cards — a multi-purpose family of card games dealing with the properties of the elements, but not specifically with the way in which they combine to form compounds. It is therefore unsuitable.

Thus, it is apparent that there are three exercises (Formulon,

Cheminoes and Compounds) which could be used in the role envisaged by our teacher. He should therefore obtain specimen or inspection copies of each of the exercises before reaching a final decision on which to buy as a class set.

EXAMPLE 2

A physics teacher in a large comprehensive school wants a 'mind-broadening' exercise to use with a class of 16 highly able sixth-formers in the period immediately after their A-level examinations. Naturally, he wants it to have a basis in physics, but his main aim is to develop data-handling, analytical decision-making, communication and debating skills. There is no tight limit on the time available, and he has up to £50 to spend.

Stage 1. Reference to Figure 6.1 directs the teacher to Figure 6.5 by answering 'no' and 'yes' respectively to the first two questions.

Stage 2. On referring to Figure 6.5, which lists exercises suitable for educating 'through' science, he finds a number of physics-based exercises of the appropriate level:

> Alternative Energy Project (p 129)
> Central Heating Project (p 131)
> Power Station Project (p 140)
> Central Heating Game (p 130)
> Power for Elaskay (p 139)
> The Power Station Game (p 144)
> Hydropower (p 134)

Stage 3. The teacher now refers to the data sheets on these exercises, and reaches the following conclusions about each.

Alternative Energy Project, Central Heating Project, Power Station Project — all turn out to be simplified, shortened versions of other exercises, with most of their 'hard' physics content removed in order to make them suitable for use with pupils who lack a rigorous physics background. They can therefore be eliminated immediately since, even if suitable for the purpose that our teacher has in mind, the original versions would be even more suitable.

Central Heating Game — a manual multi-disciplinary multi-project pack containing five projects dealing with different aspects of domestic central heating and insulation. The main physics-based project is, however, a straightforward technical case study carried out by pupils working either singly or in pairs, and is therefore not suitable for cultivating communication or debating skills.

Power for Elaskay — a manual simulated case study on alternative energy which is highly interactive and involves fairly lengthy physics-based calculations. It takes roughly three hours to complete and costs £7 for a class set. It thus appears to be ideally suited to the sort of usage our teacher has in mind.

The Power Station Game — a manual simulation game based on the planning of a large power station. It involves lengthy physics-based calculations and is highly interactive. It takes roughly two and a half days to complete and costs £25 for a class set. Again, it appears to be ideally suitable.

Hydropower — another manual multi-disciplinary multi-project pack, this time dealing with the topic of hydro-electric pumped storage. The main physics-based project again turns out to be a relatively non-interactive technical case study, however, and is therefore not really suitable for developing communication and debating skills.

It follows from the above that two exercises (Power for Elaskay and The Power Station Game) appear to be ideally suitable for the type of usage envisaged by our physics teacher. Depending on the time he wants to devote to the work, and the money he is prepared to spend, he could therefore decide to use either — or both.

Adapting exercises to meet user needs

In some cases, on working through the selection process described above, an exercise will be found that appears to be exactly suitable for the role envisaged. If such an exercise is not found, a potential user is faced with three possible courses of action. First, he can abandon the idea of using a game, simulation or case study and adopt some alternative teaching strategy. Second, he can try to find an exercise that is sufficiently close to what he wants to enable it to be used after some modification to its format, structure, content or logistics. Third, he can decide to design his own exercise. We will now examine the second of these options, leaving the third to be covered in Chapter 7.

There is one important piece of advice that the authors of this book wish to give to potential users of games and simulations. *Never be afraid to adapt an exercise to meet your particular requirements.* Remember that the educational outcomes *you* want an exercise to achieve may well be different from those the designers of the exercise had in mind. Once you have purchased a game, simulation or case study, you should feel free to make use of it *in any way you see fit;* it is, after all, simply a collection of resource materials.

In practice, there are two basic ways in which an exercise can be adapted:

1. The resource material can be used in a different way from that recommended by the designers (eg by omitting certain material from the exercise, altering the programme or time scale, or re-organizing the method by which the material is exploited).
2. The original resource material may be modified (eg by removing unwanted sections, altering or extending sections, or adding completely new sections or material).

As illustrative examples of these two different forms of adaptation, let us examine some of the ways in which Power for Elaskay has been used.

We saw in Chapter 4 that this exercise consists of a set of five parallel case studies on different alternative energy resources leading into a plenary session in which the participants have to devise a rolling programme for meeting future electricity needs on the hypothetical island of Elaskay. The exercise contains lengthy technical and economic calculations, and is primarily intended for use at upper secondary and lower tertiary levels with students who have a scientific background. It takes roughly three hours to complete.

On several occasions, the authors have run a shortened version of Power for Elaskay that omits the technical and economic calculations (eg at scientific and educational conferences, where the time available was restricted, or with groups who lacked the scientific background needed to tackle the full exercise). This was done simply by providing the participants with completed work sheets (photocopied from the teacher's guide) rather than the uncompleted work sheets supplied in the game package proper. On other occasions, the exercise has been extended by using the individual case studies and the rolling programme as 'jumping off' points for further, more detailed work. Neither of these adaptations required any modification of the original resource material (other than duplication of pages from the teacher's guide) and yet produced exercises of a radically different type from the original.

The authors have also produced a simplified, slightly abridged version of Power for Elaskay specifically for use in science in society-type courses (the Alternative Energy Project — see page 129). This was done by re-writing the various project sheets and work sheets in order to reduce the amount of technical calculations, simplifying the scenario, and completely re-organizing the logistics of the game package. In this way, it was possible to produce an exercise which could be used with a totally different target population but which retained all the essential characteristics of the original case study.

Guide to the use of gaming and simulation exercises

For convenience, the use of a game, simulation or case study in a teaching situation can be divided into three phases, namely preparatory work, the actual running of the exercise and the debriefing.

These will now be examined in turn.

(a) *The preparatory work.* For most exercises, the preparatory work required can be broken down into the following elements:
   (i) make sure that the game package is available and complete;
   (ii) make sure that all other facilities needed for running the exercise will be available as and when required (eg accommodation, audio-visual or computer facilities, extra staff);
   (iii) check that you will have a suitable number of participants to ensure that the exercise can be run effectively (if necessary, 'borrow' pupils or students from another class in order to

make up the numbers);

(iv) make yourself thoroughly familiar with the *complete* game package and, in particular, with the teacher's guide (if one is supplied);

(v) carry out any preliminary teaching that may be necessary;

(vi) issue any introductory material, and make sure that the participants know what the exercise will involve; if necessary, allocate students to roles (or groups) and issue any necessary briefing or resource material.

All these points may appear rather obvious (especially to those with some experience of running games and simulations), but they are all essential if the exercise is to run smoothly.

(b) *Running the exercise.* Unless you are deliberately departing from the method of organization recommended by the designer, the key to the successful running of a game, simulation or case study is to make sure that you follow the instructions given in the teacher's guide. If such a guide is not included in the game package, it is strongly recommended that you prepare your own.

(c) *The debriefing.* Practically all workers in the gaming and simulation field agree that a debriefing session of some sort is *essential* if full educational value is to be derived from a game, simulation or case study. The form of this debriefing will depend on the nature of the exercise involved, but should generally include the following three elements:[33]

(i) review of the actual work of the exercise, and discussion of any important points that are brought up by the participants;

(ii) discussion of the relationship between the exercise and the subject matter on which it is based (eg discussion of the degree of realism of the exercise in the case of a simulation);

(iii) discussion of any broader issues raised by the exercise.

If the exercise was itself being tested in any way (eg if the teacher was trying out a newly modified version of an existing game or simulation), a fourth element should be included, namely discussion of possible methods by which it could be improved.

The debriefing session is particularly important in the case of exercises that involve role play or place the intrinsic subject matter in a social, political, economic or environmental context; indeed, in these cases, it can be the most important part.

Chapter 7

Designing your own exercises

Since 1973, the authors have been involved in the development of over 25 educational games, simulations and case studies, for use at both secondary and tertiary levels. Despite the widely differing character of these exercises, however, it has been possible to adopt a single underlying rationale, based on a systems approach, to their design and development. This will now be described.

General description of the development process

The process by which a game, simulation or case study is developed and exploited falls naturally into three distinct phases (see Figure 7.1).

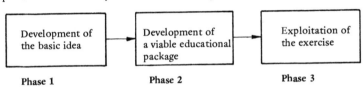

Figure 7.1: The three phases of the development process

The first of these phases is the development of the basic idea. The second (generally the longest) is the cyclical process by which a viable educational package is developed and field-tested. The third is the exploitation of the final package. Let us now examine the three phases in detail.

PHASE 1 — DEVELOPMENT OF THE BASIC IDEA

The decision to design an educational game, simulation or case study should be reached as a result of the following sequence of events:

(a) Identification of a gap in the curriculum (ie a set of specific educational outcomes it would be desirable to achieve with a specific target population).

(b) Realization that an exercise of the game, simulation or case study type could be an effective means of filling such a gap.

(c) Thorough search through the range of existing exercises with a view to finding a suitable exercise (or an exercise that could be adapted to the role envisaged) and failure to find any such exercise.

(d) Appraisal of possible alternative strategies for filling the gap in the curriculum and realization that none of these would probably be as effective as a game, simulation or case study.

Thus, to use Popperian terminology,* the starting point in the development process should be the identification of a *problem situation* — in this case, the existence of a gap in the curriculum and the absence of a suitable means of filling it. The next step should be the formulation of a *tentative solution* to the problem situation — in this case, the development of the basic idea for a suitable exercise.

As Popper has shown, the progression from problem situation to tentative solution is essentially a creative process (like the creation of a work of art or the formulation of a scientific theory) and therefore cannot be described in strictly rational, logical terms. In the case of the development of the basic idea for an exercise like an academic game, however, the creative process can be helped along by tackling the problem in a systematic manner. The approach that the authors have found useful is shown in schematic form in Figure 7.2.

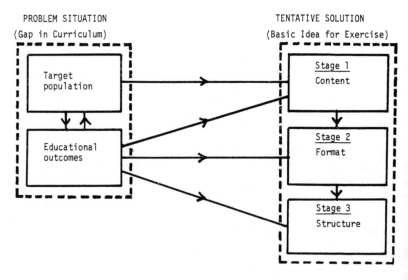

Figure 7.2: Developing the basic idea for an educational game, simulation or case study

* The philosopher Karl Popper has argued that virtually all problems can be tackled using the same basic methodological approach.[86, 87, 88] This can be summarized as

$$P_1 \longrightarrow TS \longrightarrow EE \longrightarrow P_2$$

where P_1 is the initial problem situation, TS a tentative solution, EE the process by which the solution is tested by systematic elimination of errors and P_2 is the new problem situation that emerges from the work (this is always different from P_1).

The first stage in the process is to determine (in broad terms) the content of the exercise. In many cases (particularly where the educational outcomes are mainly cognitive in nature) this will follow more or less directly from the problem situation already identified. In other cases (particularly where the desired outcomes lie mainly in the affective domain) the choice of suitable content may not be quite so straightforward. Indeed, in the latter, the content is often merely a foundation on which a structure capable of achieving the desired educational outcomes may be built (see, for example, Proteins as Human Food (page 169) and Polywater (page 167)) so that the choice of specific content need not be particularly critical.

The second stage is the choice of format (card game, board game, manual exercise* or computer-based exercise). This will depend partly on the general nature of the desired educational outcomes and partly on the specific content provisionally selected. The suitability (or otherwise) of different formats for achieving different types of educational outcomes is summarized in Figure 7.3 (see Chapters 3, 4 and 5 for more detailed information).

Type of Educational Outcomes Required \ Format	Card Game	Board Game	Manual Exercise	Computer-Based Exercise
Teaching or Reinforcing the Basic Academic Content of a Subject	Very suitable- particularly with younger age groups	Very suitable- particularly with younger age groups	Very suitable with all age groups	Very suitable, particularly with older age groups
Cultivating Broader Skills (Educating "through" Science)	Not really suitable	Can be of limited use	Very suitable	Can be quite useful - particularly with older age groups.
Teaching About Science or Technology	Not really suitable	Can be of limited use	Very suitable	Can be quite useful - particularly with older age groups

Figure 7.3: The suitability of different formats for achieving different educational outcomes

The third and final stage in the development of the basic idea for an exercise is the choice of structure. Here, the range of structures possible in any particular case will largely be determined by the specific nature of the educational outcomes required. The possible structures available within each format will now be briefly examined.

Card games. Here, a wide range of standard structures is available (eg rummy, solitaire, happy families), and the designer should work

* In the limited sense used in Chapter 4 — see page 53.

systematically through these to see if any can be adapted to the type of usage he has in mind (see Chapter 3 for typical examples).

Board games. Here, at least four basic structures are possible:

(i) use of the board as a two-dimensional matrix on which patterns may be built;
(ii) use of the board to provide a predetermined linear path (or paths) along or around which players have to progress;
(iii) use of the board as a field for mobile, two-dimensional play;
(iv) use of linear activity round the perimeter of the board to control two-dimensional activity on the interior.

Examples of how these different structures can be used to achieve specific educational outcomes are discussed in Chapter 3.

Manual exercises. Here, there are four basic structures:

(i) the simple linear structure, in which the participants work systematically through the same predetermined series of activities;
(ii) the simple radial structure, in which the participants carry out different (albeit often related) activities using different resource materials and then take part in a discussion, debate or meeting of some sort;
(iii) the composite structure, which combines linear and radial elements;
(iv) the multi-project structure, where a single set of resource materials is exploited in a number of separate projects.

In general, the simple linear structure is most suitable for exercises where a complex case study of some sort has to be broken down into manageable stages, and where the educational objectives lie mainly in the cognitive domain. The simple radial structure, on the other hand, is most useful in cases where the different points of view or starting points in a problem situation have to be examined, compared and discussed, and where the main educational objectives are non-cognitive in nature (cultivation of communication skills, inter-personal skills, desirable attitudinal traits, etc). The composite structure can be used in more complicated cases where outcomes of both types need to be achieved, but this is not recommended unless the designer has previously had some experience of developing simpler exercises. Similarly, the multi-project structure is not recommended for beginners.

Examples of exercises that use these different structures are examined in detail in Chapter 4.

Computer-based exercises. By their very nature, computer-based exercises are essentially linear in structure, although they can vary enormously in the amount of branching they employ and in the extent to which the user interacts with the program. If the exercise is to be a worthwhile *educational* experience (as opposed to a mere vehicle for imparting information),it is essential that the degree of interaction be high. The designer should therefore try to structure his exercise so that the computer is rather more than just a super-fast calculator.

The way in which the computer medium can be used to achieve different types of educational outcome is discussed in detail in Chapter 5.

PHASE 2 – DEVELOPMENT OF A VIABLE EDUCATIONAL PACKAGE

The second phase in the development of an educational game, simulation or case study involves:

(a) converting the basic idea from its original conceptual form to more tangible form (an actual package of materials), and
(b) refining the package by means of field trials designed to determine whether it is capable of doing the job for which it is intended.

(In Popperian terms, (a) is essentially a continuation of the development of the tentative solution to the original problem situation, while (b) is the next crucial step in his methodological scheme, namely the *error elimination* process.) As in the case of Phase 1, the work should be tackled in a systematic manner, using an approach of the type shown schematically in Figure 7.4.

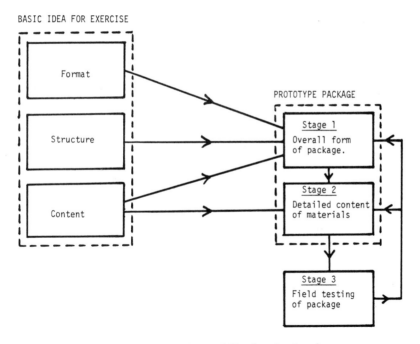

Figure 7.4: Converting the basic idea into a viable educational package

The first stage in the process is the determination of the overall form that the package is to take. To a large extent, this will depend on the type of format and overall structure provisionally selected for the exercise (see below).

Card game. Teacher's guide and/or game manual plus pack(s) of cards.
Board game. Teacher's guide and/or game manual plus board and ancillary materials (playing pieces, cards, dice, etc).
Manual exercise. Teacher's guide, introductory material and resource material (whose nature will depend strongly on the type of structure adopted).
Computer-based exercise. Teacher's guide (if appropriate), student manual(s) and actual program(s).

For further information, see Chapters 3, 4 and 5.

The second stage is the design of prototype versions of all the various materials that are to be included in the package. Here, the best advice the authors can give to a would-be designer is that he or she should study the materials contained in an existing exercise of similar format and structure to that which he or she is trying to produce and then base his or her own materials on these.

Whatever the format and structure, the primary aim should be to produce a package of materials that can be used not only by the designer but also by any other teacher or lecturer who wishes to run the exercise. For this reason, it is *essential* to include a comprehensive teacher's guide or game manual that provides sufficient information for this to be done without the user having to refer back to the designer for help or guidance. Such a document should therefore include the following elements:

— a list of the educational outcomes the exercise is designed to achieve;
— a list of the contents of the package;
— detailed instructions on how to organize the exercise.

With regard to the other resource materials to be included in the package (ie those which will be used by the participants rather than the organizer), it is important to make sure that each item:

— fits into the general context of the exercise;
— is consistent with all the other material in the package;
— is capable of fulfilling its own specific function.

This will almost certainly necessitate a certain amount of 'tuning' (ie revising or amending particular items as the work progresses in order to produce a self-consistent and balanced exercise).

The third stage in the development of a viable package is the process by which it is field tested in order to determine whether it is (a) logistically and operationally sound and (b) capable of achieving its design objectives. This corresponds to the *error elimination* phase of Popper's methodological scheme, and is possibly the most important stage in the entire development process.

The first field test of an exercise should be run by the designer himself, preferably with a group drawn from the target population for which it is intended. This can have a number of different outcomes.

(a) The exercise turns out to be a complete disaster (particularly from the educational point of view). This indicates that the whole concept should be critically re-examined, and the exercise either abandoned or drastically modified.

(b) The exercise proves successful in some respects but less so in others, revealing a number of basic (but not disastrous) organizational and/or educational flaws in its design. This is a satisfactory outcome from the designer's point of view since it

   (i) shows that his basic idea appears to be sound;
   (ii) identifies those areas where some modification is required, and
   (iii) gives some clues as to what the modifications should be.

All such changes, whether relating to the overall form of the exercise, to the detailed contents of the resource materials or to the organizational procedure, should be made before any further field tests are undertaken.

(c) The exercise proves completely successful in all respects, apart from comparatively minor points of detail which can be remedied without a major re-write. This is even more satisfactory (albeit highly unlikely!) and indicates that the exercise can be safely handed over to colleagues for further, more rigorous field testing.

Once the designer is satisfied with his exercise (ie has finally achieved outcome (c) above), he should allow it to be field tested by at least three colleagues, either in his own school or college or (preferably) in other establishments. This set of field trials is, if anything, even more important than the first, and should be conducted rigorously. Both the organizers and the participants should be actively encouraged to criticize the exercise by pointing out any aspect(s) they feel could be improved. Such constructive criticism (which lies at the very heart of Popper's methodology) should be obtained using suitably designed questionnaires and interviews. It should then be used as the basis for any further modifications of the exercise that prove necessary.

Only when criticisms have been reduced to trivia should the designer feel satisfied that he has produced a viable, educationally valuable exercise. This may take several cycles of field trials and revision.

PHASE 3 — EXPLOITATION OF THE EXERCISE

If a designer has satisfactorily completed Phase 2 above and has produced what appears to be a worthwhile educational exercise, he should seriously consider publishing his work and making his exercise generally available. All too often, potentially valuable games, simulations and case studies are used only by the designer and a few of his immediate colleagues. In many cases, such exercises would be of considerable value to the educational community at large.

A designer can publicize his work in a number of ways. First, he can write articles for educational journals, eg

— journals dealing with the teaching of science and its various branches;
— journals dealing with educational methodology;
— journals dealing specifically with the field of gaming and simulation.

Second, he can give talks at conferences, meetings of professional bodies, etc. Third, he can run demonstration sessions of his exercise for colleagues and other interested parties. The various national and international bodies formed to promote the use of gaming and simulation techniques (see the list at the end of this chapter) can be particularly useful in helping designers publicize their work.

With regard to making an exercise generally available, there are three basic ways in which this can be done:

(a) The designer can produce a do-it-yourself kit that enables other people to prepare their own copies of his or her package.
(b) He or she can produce multiple sets of the package.*
(c) The designer can approach an appropriate commercial or professional body with a view to having the exercise published and marketed on a large scale. In this case, the designer is advised to try to find an organization that has already published exercises of a similar type or in the same subject area. The data sheets given in Part 2 should be of considerable help here.

* In both cases, the designer can either give away or sell these materials, or make arrangements for them to be made available through some other body.

General hints for would-be designers

In conclusion, the authors would like to offer two general pieces of advice to potential game designers.

First, do not work on your own unless it is absolutely necessary. It is *much* easier to produce a good exercise if you can find a colleague (or group of colleagues) to help you. The interchange of ideas and mutual criticism that such teamwork makes possible will, almost without exception, not only speed up the development process but also give rise to a much better final product.

Second, remember that the human mind is like an iceberg, with only a small fraction (the conscious mind) visible and by far the largest part (the subconscious mind) hidden. It is generally accepted that the latter plays a major role in the creative process, so you should give it every chance to do so. If you are stuck for an idea, stop work on that particular problem and do something completely different; your subconscious mind will often come up with a solution, apparently quite spontaneously. This is particularly true if you think about a problem last thing at night, just before going to sleep. Try it: it often produces results.

List of principal organizations that promote academic gaming and simulation

1. ISAGA (International Simulation and Gaming Association). Contacts: *Europe* — Dr K Bruin, Ubbo Emmius, POB 1018, Leeuwarden, Netherlands; *North America* — Professor R Duke, University of Michigan, 2150 Art & Architecture Building, Ann Arbor, Michigan, USA.

2. NASAGA (North American Simulation and Gaming Association). Contact: Dr B Lawson, Barry Lawson Associates, Inc, 148 State Street, Boston, MA 02109, Mass, USA.

3. SAGSET (Society for Academic Gaming and Simulation in Education and Training). Contact: Hon Secretary, SAGSET, Centre for Extension Studies, University of Technology, Loughborough, Leics, LE11 3TU, England.

Chapter 8:

Evaluating games, simulations and case studies

Although impressive claims are often made regarding the educational value of games, simulations and case studies, the supporting evidence is nearly always lacking. On far too many occasions, a simulation or game is judged to be successful simply on the grounds that it 'works'. This, in itself, tells us little about the extent to which the overall aims and objectives have been achieved, since the only way in which such information can be acquired is by carrying out a systematic *evaluation* of the exercise in question.

While most workers agree that evaluation is important, there is less agreement as to how this evaluation should be conducted. When the exercise is a game, simulation or case study designed to achieve a range of non-cognitive outcomes, it becomes particularly difficult to devise valid and workable evaluation techniques. Thus, although gaming and simulation techniques are being increasingly used in many areas of education, research into evaluation of the effectiveness of such methods has lagged behind the development of the methods themselves.

In this chapter, we will take a look at some of the techniques that can be used to assess the effectiveness of science-based games, simulations and case studies. Before we do so, however, we will discuss some of the broader aspects of evaluation, and then examine some of the evaluation work that has actually been done on science-based games and simulations.

The evaluation debate

Perhaps we should start by answering the question 'Why should we bother to evaluate games, simulations and case studies at all?' After all, the individual teacher's own judgement on the effectiveness of the teaching methods he employs has usually been considered sufficient in the past. If the systems approach to the development of educational exercises advocated in the previous chapter is adopted, however, evaluation clearly plays a key part in the development process, since it can identify particular strengths and weaknesses of an exercise, and hence provide a firm basis for its modification and improvement. Thus, the basic question is not 'Should we evaluate?' but rather 'How should the evaluation be carried out?'

One current area of controversy in educational evaluation is concerned with the respective merits of two distinctly different approaches – the so-called *agricultural/botanical* and *social/anthropological* approaches.[89] The former is normally considered to be the 'scientific' approach to evaluation. The experiments used have tight controls, and the resulting outcomes are carefully measured using objectives-oriented evaluation procedures.

The social/anthropological approach, on the other hand, is more concerned with studying the *ongoing* process that takes place *during* an exercise. It uses far more subjective methods, and is based on the arguments that (a) the variables in educational experiments are numerous and hence cannot be readily controlled, and (b) the inputs and outputs of such experiments are not easily measurable.

Some experimenters have attempted to use the 'scientific' approach to compare the outcomes of gaming methods with more traditional methods of instruction. The value of such comparative research has, however, been widely criticized on the grounds that the 'novel' and 'traditional' methods each have particular strengths and weaknesses, and that it is not valid to compare their respective performances in achieving a set of 'compromise' objectives.[21, 33, 90, 91] The controlling of all the variables which might affect the outcome of the experiment is another major problem.

Other workers have suggested that the real educational value of games, simulations and case studies simply does not lend itself to measurement using formal evaluation procedures.[92, 93] This is echoed by Bloomer, who points out that such teaching methods very often involve complex human interactions of the type that formal input/output evaluation measurements ignore.[89]

In general, however, these broad issues are of much more interest to educationalists and educational technologists than to practising teachers. Such teachers are likely to be interested in evaluation only in so far as it can help them to improve the effectiveness of exercises that they themselves have adapted or designed. When planning such an evaluation, it is recommended that they try to adopt a middle path between the purely objective and purely subjective approaches outlined above. Whether the emphasis should tend towards the former or the latter will depend on the nature of the exercise in question and the specific educational purpose for which it is to be used.

Review of evaluation work carried out to date

The reason why 'hard' research into the effectiveness of gaming and simulation methods has not developed as far as many workers would have liked is probably a combination of several factors. These include the lack of general agreement regarding the basic approach that should be adopted, the poor quality of much of the research that has in fact been carried out, and the intrinsic difficulty of devising valid and workable assessment tools in cases where attitude and broad non-cognitive skills (eg communication, decision-making and inter-personal skills) are involved.[94] In addition, most studies of the educational outcomes of games, simulations and case studies have been strictly short-term, or have had findings that could not be generalized.

The majority of attempted evaluations of gaming materials have been carried out in America. It is apparent, however, that very few well-planned and unambiguous research efforts have been reported so far. The lack of hard evaluative evidence is used by some workers to disparage the use of gaming and simulation techniques, despite the fact that the use of more traditional teaching methods has no better empirical justification.[95]

The tentative conclusions which *can* be drawn from evaluation research have already been given in Chapter 1. In short, these are (a) that games, simulations and case studies appear to have no advantage over other methods in teaching basic facts and concepts, (b) that the methods appear to have considerable potential in achieving attitude development, and (c) that the methods strongly motivate students, who appear to enjoy participating in such exercises.

Evaluation work carried out specifically on science-based games, simulations and case studies can conveniently be sub-divided into three broad categories, namely:

1. Evaluation of the effectiveness of other people's published exercises.
2. Developmental testing, by designers, of their own exercises in order to assess their potential and identify weaknesses.
3. Large-scale evaluation of the cumulative effects of a specific programme of exercises designed to fit in with and complement an existing science curriculum.

Illustrative examples of each of these categories will now be examined.

EVALUATION OF OTHER PEOPLE'S EXERCISES

A number of evaluations of commercially available materials have been reported — generally carried out by teachers interested in obtaining feedback on their effectiveness. One example is the independent evaluation by Vaughan[96] of two of the chemistry-based card games produced by Heyden & Sons, namely Chemsyn and Element Cards (see p 151 and p 156). This study was done on a relatively small scale with one class of students. The evaluation involved testing the students before and after their experience with the card games in order to determine their ability to carry out the operations used in the games and to test their knowledge of the information provided on the cards. The experimental group was compared with a control group. (It should be emphasized that, in this case, it was the *card games* (and not the *students*) that were being evaluated.) Vaughan concluded that his tests indicated improved knowledge of the specific chemical principles under investigation, although, as is so often the case, he added the rider that further evaluations were needed with larger groups in order to establish the statistical significance of these observations.

Another interesting example of this broad category is the evaluation of The Power Station Game (see p 144) that was carried out by Millar[54] in Australia. The exercise was used in a training course for primary teachers — a use which the designers of the exercise had not envisaged. The evaluation was again on a relatively small scale. It was effected via a series of short questionnaires that the students filled in at various stages of the exercise, giving their current rating of the game in terms of enjoyment, complexity, relevance and content. The evaluation showed the exercise to have been a success (in the opinion of the evaluator), and has since been repeated each time the exercise has been used in order to provide feedback.

DEVELOPMENTAL TESTING OF PROTOTYPE MATERIALS

This is probably the area in which most evaluation work is done, although the results are not always reported. As stated earlier, if one adopts a systematic approach to the design of a game, simulation or case study, evaluation of the package during its development is *vital*.

A trial of an exercise may simply be a field test whose main object is to identify design faults and ambiguities, the information thus gained being noted, and the exercise duly adapted. This type of evaluation is basically part of the final 'tuning' process for the exercise (see Chapter 7). Alternatively, the trial may go further, and involve a formal evaluation of the effectiveness of the exercise in achieving its educational design objectives.

This type of developmental testing has been described for three chemistry-based exercises, namely Proteins as Human Food, What Happens When The Gas Runs Out? and Polywater[10] (see p 168, p 177 and p 167). All these exercises are designed to develop both cognitive and broader non-cognitive skills. Because of this, a variety of assessment tools was used, depending on the type of outcome currently under scrutiny. These included multiple-choice questions on the cognitive content, rating scales to assess students' confidence in various communication skills, attitude scales, student interviews, and recorded observation of the playing sessions. The use of such a battery of techniques was necessary because of the wide range of outcomes being investigated.

LARGE-SCALE EVALUATION OF A PROGRAMME OF EXERCISES

Although the types of evaluations discussed above play a vital part in assessing the value and effectiveness of individual exercises, the generalizing power of such research is usually limited to the specific conditions under which the research is conducted. In most cases, the general validity is restricted by the small number of participants, and also by the fact that the designer of the exercise runs the playing session and organizes the research himself. The broader questions raised earlier in this chapter regarding the general applicability and limitations of games, simulations and case studies can only be answered by well-planned, large-scale research programmes in which the designer of the package is *not* directly involved in the administration of its use.

One such large-scale evaluation of the cumulative effect of a series of games, simulations and case studies designed to fit in with an existing science curriculum has been carried out by Norman Reid, working from the University of Glasgow.[65] Reid was interested in developing a 'social awareness of science' in 13 to 14 year old school pupils. The subject area was limited to the content of the Scottish O-grade chemistry syllabus and, for the purpose of the study, five main areas of interest were chosen from the existing syllabus. These were:

1. Historical implications of the development of chemistry.
2. Domestic implications of chemistry.
3. Industrial implications of chemistry.
4. Economic implications of chemistry.
5. Socio-moral implications of chemistry.

To help pupil development in these areas, a series of 12 interactive teaching packages was constructed, most of which involved extensive use of game/simulation/case study techniques. These packages were designed to use the basic content of the syllabus as a vehicle for non-cognitive development.

Each package was initially tested with over 250 pupils and ten teachers from eight schools in order to ensure that it was workable in terms of vocabulary, content, time and clarity. At this level of testing, most teachers and pupils found the packages acceptable and useful.

Reid then investigated whether the packages achieved any objectives from the five broad areas in which he was interested *in the long term*. To this end, a seven-month experiment was set up. Several hundred third-year pupils following O-grade chemistry courses in schools throughout Scotland completed a minimum of four packages over a six-month period, the packages being chosen to fit in with their usual course work. At the end of seven months, they were assessed, and the results compared with those obtained in a second assessment carried out with a similar number of pupils who had not used any of the packages. A parallel experiment was undertaken with fourth-year pupils to provide corroboration and comparisons.

The assessment was carried out using questionnaire booklets in which pupils answered roughly 75 questions in the space of 30 minutes. The evaluation items were validated by sample interviews with the pupils. Obviously, a wide range of information was obtained from these assessments, but the degree of consistency between the two experiments was remarkable. A summary of how the experimental pupils (who had used the packages) compared with the control pupils (who had not) in each of the five areas is given below.

Historical — there was insufficient historical emphasis in the packages, but some good specific achievements.

Domestic — the response was encouraging, although some gaps were apparent.

Industrial — the results were remarkably good, indicating that there was much more scope for development in this area.

Economic — there was evidence of consistently good achievements, with almost every facet being developed to some extent.

Socio-moral – impact appeared to be limited in this area, especially when objectives became more personal.

Reid concluded that non-cognitive outcomes *can* be achieved with O-grade chemistry pupils in the area of social awareness through the use of interactive teaching materials. Pupil reaction was very enthusiastic, and teacher reaction varied from initially sceptical to enthusiastic, becoming progressively more favourable with experience. Student non-cognitive achievement was relatively permanent in that it lasted for several months. Reid claims that the results of his research have a wide applicability to other subject areas, and points out that the scope for research into the development of the various non-cognitive outcomes of a science education is enormous.

Some evaluation techniques

Let us now examine the various techniques that can be used to evaluate the effectiveness of games, simulations and case studies in science education.

COGNITIVE TESTS

The assessment procedures traditionally used in education are concerned almost entirely with measuring *cognitive* attainment (eg knowledge gained). Such procedures include essay-type questions, objective items and unique answer questions. In terms of reliability, validity, ease of application and statistical treatment, the last two question types appear well suited to measuring purely cognitive gains from games, simulations or case studies. The actual experimental design adopted will depend on several factors, including the number of participants available and the possibility of obtaining suitable control groups.

NON-COGNITIVE TESTS

A major problem arises when any attempt is made to measure the development of attitudes and skills which are not purely cognitive, because traditional assessment methods are often inappropriate. The suitability of assessment techniques is further restricted by the need to have procedures which are simple in application, are acceptable to students, are as valid and reliable as possible, and produce results that can be meaningfully analysed.

Such conditions limit the choice to methods which can be broadly classified under two headings: *observational techniques* and *self-reporting techniques*.

Observational techniques
When certain attitudinal objectives are involved, or when the development of communication or inter-personal skills is a major aim of the exercise, direct observation of individual or group behaviour can be of use in

monitoring its effectiveness. Closed-circuit television or a simple audiotape recording of the activities in a group session can act as a permanent record of the exercise so that subsequent analysis can be attempted. This analysis can take the form of simply cataloguing the numbers and patterns of communication within a group. At the other extreme, it can involve some sort of 'interaction process analysis', which may provide an insight into the social psychology of the group and the personalities of individual members.

Self-reporting techniques
Several such techniques have been devised, the most important of which will now be described.

(a) *Likert scales*[97] are commonly used in attitude measurement, and appear deceptively easy to devise and administer. There is, however, a lot of skill required in producing good Likert items which (i) are valid, and (ii) provide good discrimination between people with different opinions on a topic. Essentially, a Likert rating scale consists of a list of statements, the person responding having to make a judgement on every statement — usually by selecting from a number of degrees of agreement and disagreement. Figure 8.1 shows a series of Likert items designed to assess students' attitudes and confidence towards group discussion in class. Such an evaluation could be used in conjunction with a simulation, game or case study designed to increase inter-personal and communication skills.

The number of options on the scale will depend on the requirements of the setter. It can be odd or even, the latter having the advantage that it does not allow participants to take refuge in a completely neutral category. Although relatively common practice, it is statistically unwise to add together scores for separate statements on a Likert scale to obtain an *overall* attitude score. This may only be done if the statements are measuring the same attitude dimension. It should, however, be possible to use Likert-type scales to identify variations in attitude and opinions within a group. This can be done by comparing participants' responses to *individual* statements in a test administered before and after a teaching unit designed to modify such opinions, or by comparing them with a carefully matched control group.

(b) The *semantic differential technique*[98] has been used as a tool for various types of assessment designed to measure the connotations that particular concepts have for individuals. Word pairs of antonyms such as 'valuable/worthless' and 'interesting/boring' are joined by a 3, 4, 5, 6 or 7 point scale. The method is based on the premiss that the word pairs are opposites, although this assumption may not always be valid because of the fact that particular words sometimes have different meanings for different students.

Figure 8.2 gives an example of a semantic differential grid used in the evaluation of Fluoridation? (see p 194).[82] The grid was used to assess students' attitudes to fluoridation before and after participating in the exercise.

Put a letter in the appropriate box to indicate your <u>personal</u> opinion about each of the following statements.

> Mark: A - if you strongly agree with the statement
>   B - if you agree with the statement
>   C - if you neither agree not disagree, or you don't know
>   D - if you disagree with the statement
>   E - if you strongly disagree with the statement.

i)   If I knew a piece of information of relevance to a class discussion I would be eager and willing to communicate it to the class.

ii) I find it easy to differentiate between facts and opinions

iii) I have confidence in my ability to argue a point during a class discussion

iv) I would accept the opinions of scientists on matters in which I am not an expert

v)   The thought of presenting a short talk to my class makes me feel very uncomfortable

vi) I benefit more from reading a text book than from actively participating in a class discussion

Figure 8.1: An example of a Likert scale

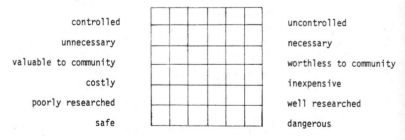

Fluoridation of Public Water Supplies

controlled	uncontrolled
unnecessary	necessary
valuable to community	worthless to community
costly	inexpensive
poorly researched	well researched
safe	dangerous

Figure 8.2: An example of a semantic differential grid

A major advantage of the method is that the items are fairly easy to write and mark. It has, however, been pointed out that the arbitrary labelling of rating scales (eg −3 to +3 or 1 to 5) is statistically groundless.[39] Such scores provide only an *indication* of the relative strengths of opinions or attitudes between different people. Statistically, it is more correct to compare differences in the actual numbers of participants responding to individual items on the scale.

(c) *Rating objectives* by means of a grid.[99] Here the objectives of an exercise are listed, and the student is asked to indicate whether each objective has been 'well achieved' through to 'not achieved at all'.

Figure 8.3 gives an example of a grid which was used as part of the evaluation of the exercise Proteins as Human Food. Such a grid would be given to students immediately after completing an exercise.

The rating is generally done using a five-point scale, although variations are possible. This method is particularly useful in cases where no other suitable technique exists for assessing the achievement of certain objectives, or as a check on other evaluation methods.

(d) *Situational techniques*[39, 100] have been used to assess the modes of thought and scientific approach of students, as opposed to their attitudes *to* science. In such an assessment, the students are confronted with a problem which in itself need have no scientific content, but which requires them to think 'scientifically' in order to solve it. Situations of this type involve students using the 'broader skills' associated with a science education (eg problem-solving and decision-making skills), and rely little on previously learned factual knowledge. Figure 8.4 gives an example of a situational technique used to assess the students' ability to control variables.[39]

Although comparatively little work has been done on the use of the technique, it would appear to have a reasonably high validity compared with some of the more artificial assessment procedures discussed above. The method is, however, sometimes cumbersome, and questions may be difficult and time-consuming to construct and apply.

(e) *Interviews* with students have sometimes been used to assess the effectiveness of games, simulations and case studies. A well-constructed interview procedure is, however, not easy to devise, administer and analyse effectively, although it would appear that the validity of the method is high.

The above list of assessment methods is by no means exhaustive, but it does give examples of some of the techniques which evaluators of science-based games, simulations and case studies have found useful. They vary quite markedly in ease of construction and validity. Indeed, an inverse relationship appears to exist between these two factors, with semantic differential and rating objectives techniques being easiest to construct but probably least valid, and interviews and situational methods being hardest to construct yet most valid if done well.

As a result, each technique has certain drawbacks, and no single technique is the best in all situations. In carrying out an evaluation, it is

At the start of this exercise you were given a list of 15 objectives. Please indicate (by means of a tick in the appropriate box) how successful or otherwise, in your opinion, the exercise has been in achieving each objective.

OBJECTIVE	VERY SUCCESSFUL	FAIRLY SUCCESSFUL	AVERAGE	NOT VERY SUCCESSFUL	DISMAL FAILURE
1. To write down the general formula for an α-amino acid					
2. To appreciate the importance of proteins in living systems					
3. To state why certain amino acids are called essential					
4. To describe a peptide link and its importance in protein chemistry					
5. To identify factors contributing to the world protein problem					
6. To extract general trends and relevant information from tabulated data.					
7. To communicate information fluently and grammatically					
8. To argue logically and precisely					
9. To teach others in public					
10. To apply a scientific approach to general problems					
11. To appreciate the value of group discussion for co-operation in discovery					
12. To identify problems and devise means of solving them					
13. To look for best compromises in matters of conflicting interest					
14. To realise the existence of interconnections across subject barriers					
15. To be aware of the limitations of science in solving some problems					

Six test runs were made on the same motor-cycle to find out the relative importance of *four* factors when economical performance is required, ie the *best* miles per gallon (mpg). From the following results, place the four factors (Type of Petrol, Average Speed, Town or Country Driving and the Experience of the Driver) in the order in which you think they are important in giving the *best* mpg. Give the most important factor a letter A in the box provided, the second best a letter B, third C and fourth D.

Test Run No.	Type of Petrol (cost/gallon)	Average Speed on Journey (m.p.h.)	Town or Country Driving	Experienced or Learner Driver	m.p.g. obtained
1	38p	35	Mainly Country	Experienced	125
2	38p	60	Mainly Country	Learner	90
3	38p	60	Mainly Country	Experienced	95
4	38p	35	Mainly Country	Learner	120
5	38p	35	Mainly Town	Experienced	105
6	32p	35	Mainly Town	Experienced	105

Factor	Importance
Type of Petrol	
Average Speed	
Town/Country driving	
Experienced/Learner Driver	

Figure 8.4: An example of a situational item

therefore advisable to use a battery of techniques, using alternative methods as a check on others.

Analysis of results

Much research in which rating scales are used can be criticized on the grounds of invalid statistics. For example, some researchers have put arbitrary numbers on scales, added up totals from the various items, and arrived at a 'score' expressing the students' attitudes to the subject in question. This is unreasonable, because the individual items will not necessarily be measuring the same parameter. It seems obvious, therefore, that items must be analysed individually. Several methods have been used for this purpose, including simple statistical analysis (eg chi-square), graphical methods or even presentation of raw data in cases where small numbers or other constraints invalidate statistical treatment of results.

While it is important that overall trends should be identified, it is equally important that individual measurements which go against the general trend should be recognized and explained. For example, it is quite possible that an exercise which is designed to develop communication skills may well have an overall beneficial effect but, for one or two participants, may actually prove inhibiting.

Conclusion

Despite the various difficulties and pitfalls described above, evaluation is an essential and valuable component of the process by which a game, simulation or case study is developed or adapted to do a particular job. The science teacher or lecturer who wishes to use such exercises in his or her courses is mainly interested in finding out whether a particular exercise will work with *his or her* students in *his or her* classroom, and whether it is capable of achieving the various educational aims in which *he or she* is interested. The best way to answer these questions is to carry out a well-planned evaluation, the results of which can then form the basis of any necessary modifications to the exercise, thus helping to ensure that its aims are better achieved next time it is used. We hope that the information given in this chapter will help make this task easier.

Part 2:
Data sheets on science-based exercises

Section 1: Physics-based exercises

ALTERNATIVE ENERGY PROJECT
Authors: H I Ellington & E Addinall
Publisher: The Association for Science Education, College Lane, Hatfield, Herts, AL10 9AA
First published: 1980
Type of exercise: manual simulated case study.
Courses for which exercise is suitable: The exercise was specially developed for use in the ASE's Science in Society AO level course, and is suitable for use in science in society and general studies courses at upper secondary and lower tertiary levels.
Educational aims: to make the participants aware of the technical and economic feasibility of exploiting different alternative energy resources; to cultivate interpretative, analytical, decision-making, communication and inter-personal skills.
Number of participants needed: The exercise is designed for use with a class of up to 20 pupils or students, but can be used equally effectively with lower numbers (min 5).
Time required: 3 or 4 school periods of roughly 35-40 minutes each or equivalent.
Contents of package: teacher's guide, student booklets.
Special facilities required: calculators, overhead projector, blank OHP transparencies and felt pens.
Outline of exercise: The Alternative Energy Project is a simplified version of Power for Elaskay (see page 139). The participants have first to carry out technical and economic appraisals of the different energy resources available on the hypothetical island of Elaskay, and then develop a 50-year rolling programme for meeting the island's electricity requirements by making use of these resources.

CAPACITOR DISCHARGE
Authors: D C F Chaundy & R D Masterton
Publisher: Schools Council, distributed by Edward Arnold (Publishers) Ltd, London
First published: 1978
Price: part of a pack Computers in the Physics Curriculum, edited by R D Masterton and D C F Chaundy, price £14.00

Type of exercise: computer simulation.
Computer language: BASIC.
Courses for which exercise is suitable: physics courses at upper secondary and lower tertiary levels.
Educational aims: to assist students' understanding of the concept of exponential decay.
Number of participants needed: The exercise can be used either individually or with small groups.
Contents of package: teacher's notes, students' leaflets, computer program.
Special facilities required: interactive computer terminal or microcomputer.
Outline of exercise: This exercise combines laboratory experiments with computations involving simple capacitor circuits.

CENTRAL HEATING GAME

Authors: A Cowking, H I Ellington, E Addinall & N H Langton
Publisher: The Institution of Electrical Engineers, Savoy Place, London WC2R 0BL
First published: 1978
Price: £9 (inc p & p)
Type of exercise: a manual multi-project pack based on a simulated case study.
Courses for which exercise is suitable: physics, engineering science, science in society, architecture, economics and general studies courses at upper secondary and lower tertiary levels.
Educational aims: to make the participants aware of the technical principles and economics of domestic central heating and insulation; to cultivate interpretative, analytical, decision-making and communication skills.
Number of participants needed: Each project is suitable for use with a class of up to 24 pupils or students, but can also be used with much smaller numbers.
Time required: The projects vary in length from 1½ - 3 hours.
Contents of package: teacher's guide, introductory booklets, project sheets and data sheets.
Special facilities required: calculators, overhead projector, blank OHP transparencies and felt pens.
Outline of exercise: The Central Heating Game is a multi-disciplinary multi-project pack, a new type of educational package that exploits the same basic set of resource materials in a series of projects designed for use in the teaching of a wide range of academic subjects (see above). It contains five basic projects, each of which deals with a different aspect of domestic central heating. All five projects are based on a simulated case study that involves calculating the heating requirements of a hypothetical bungalow and examining the cost effectiveness of different central heating systems and insulation measures.

CENTRAL HEATING PROJECT
Authors: H I Ellington & E Addinall
Publisher: The Association for Science Education, College Lane, Hatfield, Herts AL10 9AA
First published: 1980
Type of exercise: a manual multi-project pack based on a simulated case study.
Courses for which exercise is suitable: The exercise was specially developed for use in the ASE's Science in Society AO level course, and is suitable for use in science in society and general studies courses at upper secondary and lower tertiary levels.
Educational aims: to make the participants aware of the principles and economics of domestic central heating and insulation; to cultivate interpretative, decision-making and communication skills.
Number of participants needed: Each project is suitable for use with a class of up to 20 pupils or students but can be used equally effectively with smaller numbers.
Time required: Each project requires 2 school periods of roughly 35 - 40 minutes each or equivalent.
Contents of package: teacher's guide, student booklets.
Special facilities required: overhead projector; blank OHP transparencies and felt pens.
Outline of exercise: The Central Heating Project is a greatly simplified adaptation of the Central Heating Game (see page 130). Like the latter, it takes the form of a multi-project pack and is based on a simulated case study on domestic central heating and insulation. It contains three basic projects, dealing with different aspects of the central heating and insulation of five different types of house. It involves no technical calculations, and is therefore suitable for use with students who lack a scientific or technical background.

CIRCUITRON
Authors: J Megarry & M Roebuck
Publisher: Circuitron was originally published by Griffin and George Ltd in 1972, but is now out of print. It should soon be available from a new publisher. For details, contact Miss J Megarry, Scottish Microelectronics Development Programme, 74 Victoria Crescent Road, Glasgow G12 9JN
Type of exercise: board game.
Courses for which exercise is suitable: The exercise is designed for use in teaching electrical circuit theory at a wide range of levels from late primary through secondary to early tertiary. It can be used in general science, physics, electronics and electrical engineering courses.
Educational aims: to consolidate and reinforce new ideas about electric circuits just discovered in the laboratory or classroom.
Number of participants needed: The game package is designed for use by up to 4 people and, by using multiple sets of the package, the exercise can be used with a class of any size.

Time required: Each game in the package requires a minimum of 10 - 15 minutes, plus an initial familiarization time of 10 - 30 minutes.
Contents of package: board, set of playing pieces, game manual (for use by teacher or students).
Special facilities required: none.
Outline of exercise: Circuitron is not a single game, but a family of games ranging from very simple to very difficult. In each game, a player (or pair of players) draws a 'hand' of playing pieces (cardboard rectangles representing electrical components such as cells, switches and bulbs) and tries to fit these into slots in the board in such a way as to make a valid electrical circuit. Each type of playing piece has an assigned value, and points are scored according to the total value of the pieces in the circuit formed. The difficulty of the game is controlled by (a) varying the number and type of pieces used and (b) varying the detailed rules by which circuits can be formed.

ENERGY CONVERSIONS
Authors: R D Stamp & W Harrison
Publisher: Longman Group Ltd, Resources Unit, 9/11 The Shambles, York
First published: 1975
Price: £0.55
Type of exercise: board game.
Courses for which exercise is suitable: physics courses at middle to upper secondary school level.
Educational aims: to make the participants aware of the importance of the sun as an energy source (directly and indirectly) on earth, and of the inter-relationships between different forms of energy such as mechanical, heat, radiant, chemical and electrical; in addition the exercise is designed to develop literary skills.
Number of participants needed: 6 players working in pairs.
Time required: 1 school double period of roughly 1¼ - 1½ hours.
Contents of package: A4 booklet in which the three outer sheets (of thin card) form the two playing boards and other game materials. The inner sheets (of paper) constitute an 8-page background reader/instructional booklet for the pupils.
Special facilities required: scissors, 60 pieces of card on which to write questions, felt pens in 6 different colours.
Outline of exercise: Energy Conversions is played in 2 parts. In the first, each pair of players initially prepares 15 - 20 questions which relate to their energy class (food and solar energy, fuels and air and water). In addition each player prepares 2 questions relating to nuclear energy. Each player has then to progress along board 1 to an energy class different from the one from which he started, by answering the questions set by colleagues. In the second (using board 2), each player has to move around the board, which represents the different energy interconversion processes, collecting tokens on the way.

FLOW THROUGH A NOZZLE
Inquiries: R Lewis, Educational Computing Section, Chelsea College,
Pulton Place, London SW6 5PR
First published: 1977
Type of exercise: computer simulation.
Computer language: BASIC or FORTRAN.
Courses for which exercise is suitable: physics courses at tertiary level.
Educational aims: to reinforce teaching on flow through nozzles.
Number of participants needed: The exercise can be used either
individually or with small groups.
Time required: 1 hour.
Contents of package: teacher's guide, students' notes, computer program.
Special facilities required: graphics terminal.
Outline of exercise: The student specifies the exact pressure of a de Lavel
nozzle. The program displays the nozzle profile beneath a plot of pressure
against position along the nozzle, and also any shock waves which occur.

FREE FALL WITH AIR RESISTANCE
Inquiries: R Lewis, Educational Computing Section, Chelsea College,
Pulton Place, London SW6 5PR
First published: 1977
Type of exercise: computer simulation.
Computer language: BASIC or FORTRAN.
Courses for which exercise is suitable: physics courses at tertiary level.
Educational aims: to allow students to study the motion of a body falling
under gravitational forces in a resisting medium.
Number of participants needed: The exercise can be used either
individually or with small groups.
Time required: 1 hour.
Contents of package: teacher's guide, students' notes, computer program.
Special facilities required: graphics terminal.
Outline of exercise: The first part of the program allows students to plot
velocity-time and distance-time graphs for different initial and terminal
velocities, and different power laws of air resistance. The second part
simulates a parachute jump and the user can search for the time of opening
needed in order to reach the ground safely in a minimum time.

GASEOUS DIFFUSION
Authors: R D Masterton, J Harris & R Lewis
Publisher: Schools Council, distributed by Edward Arnold (Publishers) Ltd,
London
First published: 1978
Price: part of a pack Computers in the Physics Curriculum, edited by
R D Masterton and D C F Chaundy, price £14.00
Type of exercise: computer simulation.
Computer language: BASIC.
Courses for which exercise is suitable: physics courses at upper secondary
and lower tertiary levels.

Educational aims: to enable study in depth of (a) why the molecules of a diffusing gas always spread out, and (b) why the molecules of a diffusing gas will never come together in one place.
Number of participants needed: The exercise can be used either individually or with small groups.
Contents of package: teacher's notes, students' notes, computer program.
Special facilities required: interactive computer terminal or microcomputer.
Outline of exercise: The exercise involves an extended study of simple models of gaseous diffusion, progressing from shuffling counters at random to an understanding of the statistics of 1 million particles moving at random between 2 halves of a box.

GRAVITATIONAL FIELDS
Author: J Harris
Publisher: Schools Council, distributed by Edward Arnold (Publishers) Ltd, London
First published: 1978
Price: part of a pack Computers in the Physics Curriculum, edited by R D Masterton and D C F Chaundy, price £14.00
Type of exercise: computer simulation.
Computer language: BASIC.
Courses for which exercise is suitable: physics courses at middle secondary level.
Educational aims: to improve students' understanding of what is meant by escape velocity.
Number of participants needed: The exercise can be used either individually or with small groups.
Contents of package: teacher's notes, students' notes, computer program.
Special facilities required: interactive computer terminal or microcomputer.
Outline of exercise: This simulation involves an investigation of simple gravitational motion, and deals with the maximum distance travelled by objects launched vertically. It attempts to teach the meaning of the term 'escape velocity'.

HYDROPOWER
Authors: H I Ellington & E Addinall
Publisher: The Institution of Electrical Engineers, Savoy Place, London WC2R 0BL
First published: 1977
Price: £15 (inc p & p)
Type of exercise: a manual multi-project pack based on a simulated case study.
Courses for which exercise is suitable: physics, engineering science, science in society, geography, economics and general studies courses at upper secondary and lower tertiary levels.
Educational aims: to make the participants aware of the technical principles and economics of hydro-electric pumped storage; to cultivate interpretative, analytical, decision-making, communication and

inter-personal skills.

Number of participants needed: Each project is suitable for use with a class of up to 24 pupils or students, but can also be used with much smaller numbers.

Time required: The projects vary in length from 1 - 3 hours.

Contents of package: teacher's guide, introductory leaflets, background leaflets, maps, data sheets and project sheets.

Special facilities required: calculators.

Outline of exercise: Hydropower is a multi-disciplinary multi-project pack, a new type of educational package that exploits the same basic set of resource materials in a series of projects designed for use in the teaching of a wide range of academic subjects (see above). It is based on a scenario that involves finding the best site for a 1000 MW hydro-electric pumped storage scheme in a hypothetical area near the west coast of Scotland and carrying out a detailed design study on each of the possible schemes. The scenario was used as the basis of the Hydropower 77 competition run by the North of Scotland Hydro-Electric Board for Scottish secondary schools in 1976-77. The package contains 6 basic projects, each of which deals with a different aspect of the scenario.

INTERP (UNIT ON WAVE SUPERPOSITION)

Author: J Harris

Publisher: Edward Arnold (Publishers) Ltd, London (Chelsea Science Simulation Project)

First published: 1977

Price: £7.25

Type of exercise: computer simulation.

Computer language: BASIC.

Courses for which exercise is suitable: physics courses at upper secondary and lower tertiary levels.

Educational aims: to focus students' attention on the physical model used to explain 'interference and diffraction observations in a way that, because of the mathematics involved, is not easy to do without a computer; to encourage a more critical attitude to the use of models in physics in general by emphasizing and investigating some of the assumptions made in this example'.

Number of participants needed: The exercise can be used either individually or with small groups.

Time required: 1½ - 2 hours.

Contents of package: teacher's guide, students' notes, computer program.

Special facilities required: interactive computer terminal or microcomputer.

Outline of exercise: The computer program calculates the intensity due to the superposition of radiation from two sources, or two slits. The program can be used for three investigations with an optional extension, in which students can use a more complex model to investigate the effects of the direction and distance factors, and of the number of secondary sources in each slit.

KEEPING WARM
Authors: R D Stamp & W Harrison
Publisher: Longman Group Ltd, Resources Unit, 9/11 The Shambles, York
First published: 1975
Price: £0.55
Type of exercise: board game.
Courses for which exercise is suitable: physics courses at upper secondary level.
Educational aims: to help the participants understand the way in which heat losses from a house can be determined and how these heat losses can be reduced by introducing a range of insulation measures.
Number of participants needed: 2 - 4 players.
Time required: 1 school double period of roughly 1¼ - 1½ hours.
Contents of package: A4 booklet whose outer sheets (of thin card) are used to prepare the board and other playing materials. The inner sheets (of paper) constitute an 8-page background reader/instructional booklet for the pupils.
Special facilities required: scissors to prepare playing materials, matchsticks for mounting spinners, electronic calculators.
Outline of exercise: The board represents a house which has to have its heating requirements determined (the areas of possible heat loss being identified). The participants have to determine (a) the heat losses and (b) sites of heat sources and the heating requirements. They then look into the advantages of insulation measures. Finally, once their system is complete, the reverse of the game board is used to complete a fuel cost and U-value analysis.

MASS SPECTROMETER
Author: R D Masterton
Publisher: Schools Council, distributed by Edward Arnold (Publishers) Ltd, London
First published: 1978
Price: part of a pack Computers in the Physics Curriculum, edited by R D Masterton and D C F Chaundy, price £14.00
Type of exercise: computer simulation.
Computer language: BASIC.
Courses for which exercise is suitable: physics courses at upper secondary and undergraduate levels.
Educational aims: to allow students to acquire an understanding of the physics of the mass spectrometer in terms of the instrument itself, methods of obtaining useful output and interpretation of output.
Number of participants needed: The exercise can be used either individually or with small groups.
Contents of package: teacher's notes, students' leaflets, computer program.
Special facilities required: interactive computer terminal or microcomputer.
Outline of exercise: The exercise is a simulation of a mass spectrometer

experiment which leads students towards the eventual identification of ions in an 'unknown sample'. Investigations from a range of supplied samples may be tried, and further samples can be added to the simulation.

NEWTON (UNIT ON SATELLITE ORBITS)

Author: J Harris
Publisher: Edward Arnold (Publishers) Ltd, London (Chelsea Science Simulation Project)
First published: 1975
Price: £7.25
Type of exercise: computer simulation.
Computer language: BASIC.
Courses for which exercise is suitable: physics courses at middle and upper secondary levels.
Educational aims: to extend students' knowledge of projectile motion; to appreciate how the application of Newton's second law and his law of gravitation lead to the prediction of satellite orbits; to create an awareness of the possible shapes of orbits; to apply the idea of conservation of energy in a new situation.
Number of participants needed: The exercise can be used either individually or with small groups.
Time required: 1½ - 2 hours.
Contents of package: teacher's guide, students' notes, computer program.
Special facilities required: interactive computer terminal or microcomputer.
Outline of exercise: The computer program uses an iterative method to calculate the path of a projectile launched horizontally. An introductory text and a series of questions prepare the student for work at the terminal. The student is challenged to find the initial velocity needed for the minimum (circular) orbit.

PBARR

Author: C Aust (available from the author at the School of Physics, Robert Gordon's Institute of Technology, Aberdeen)
First published: 1979
Type of exercise: computer simulation.
Computer language: BASIC.
Courses for which exercise is suitable: physics, chemistry and engineering courses at upper tertiary level.
Educational aims: to enable the student to investigate the phenomenon of quantum-mechanical penetration of a potential barrier by a beam of particles.
Number of participants needed: 1 or 2.
Time required: roughly 30 minutes.
Contents of package: students' notes, computer program.
Special facilities required: interactive computer terminal, pocket calculator.
Outline of exercise: The program requests the height and width of the barrier and the mass and energy of the particles. It responds by printing

out the probability of penetration of the barrier by the particles. The effect of varying each or all of the parameters can be examined.

PHOTO-ELECTRIC EFFECT
Authors: M Doss & R D Masterton
Publisher: Schools Council, distributed by Edward Arnold (Publishers) Ltd, London
First published: 1978
Price: part of a pack Computers in the Physics Curriculum, edited by R D Masterton and D C F Chaundy, price £14.00
Type of exercise: computer simulation.
Computer language: BASIC.
Courses for which exercise is suitable: physics courses at upper secondary and lower tertiary levels.
Educational aims: to provide students with simple practical experience of photo-electric emission; to enable students to understand the development of Einstein's photo-electric equation; to enable students to understand how Millikan's photo-electric experiment was used to confirm Einstein's equation.
Number of participants needed: The exercise can be used either individually or with small groups.
Contents of package: teacher's notes, students' leaflets, computer program.
Special facilities required: interactive computer terminal or microcomputer.
Outline of exercise: This computer exercise is designed to be used in conjunction with a series of more formal laboratory experiments. It involves a simulation of the photo-electric effect followed by a simulation of Millikan's apparatus.

PHYSICS CROSSWORDS
Authors: A G Hudson & C M Eveling
Publisher: Sigma Technical Press, 23 Dippons Mill Close, Tettenhall, Wolverhampton WV6 8HH
First published: 1978
Price: £1.25
Type of exercise: set of crossword puzzles.
Courses for which exercise is suitable: physics courses at middle and upper secondary level (specifically designed for English O- and A-level courses).
Educational aims: to serve as a reinforcement and revision tool covering all aspects of physics included in the above courses.
Number of participants needed: can be used with groups of any size or for private study.
Time required: Each crossword requires roughly 30 minutes to complete.
Contents of package: teacher's guide, set of 20 crosswords (1 copy of each) with solutions.
Special facilities required: none.
Outline of exercise: The participants complete the crosswords in the conventional way. The 20 puzzles in the pack are designed for use at

different levels and cover different branches of physics (heat, light, electricity, etc).

PLANETARY MOTION

Authors: D C F Chaundy & R D Masterton
Publisher: Schools Council, distributed by Edward Arnold (Publishers) Ltd, London
First published: 1978
Price: part of a pack Computers in the Physics Curriculum, edited by R D Masterton and D C F Chaundy, price £14.00
Type of exercise: computer simulation.
Computer language: BASIC.
Courses for which exercise is suitable: physics courses at middle and upper secondary levels.
Educational aims: to assist in the teaching of elementary ideas of astronomy, simple relationships between variables, and Newton's law of gravitation and its applications.
Number of participants needed: The exercise can be used either individually or with small groups.
Contents of package: teacher's notes, students' notes, computer program.
Special facilities required: interactive computer terminal or microcomputer.
Outline of exercise: Students analyse real and generated orbit data which lead them towards the laws of planetary motion. At the younger level, this exercise is suitable for students who are learning about direct proportionality, while for older students the exercise also covers simple Newtonian gravitation applied to orbit theory.

POWER FOR ELASKAY

Authors: H I Ellington & E Addinall
Publisher: The Institution of Electrical Engineers, Savoy Place, London WC2R 0BL
First published: 1978
Price: £7.00 (inc p & p)
Type of exercise: a structured lesson built round a (manual) simulated case study.
Courses for which exercise is suitable: physics, engineering science, science in society and general studies courses at upper secondary and lower tertiary levels.
Educational aims: to make the participants aware of the technical and economic feasibility of exploiting different alternative energy resources; to cultivate interpretative, analytical, decision-making, communication and inter-personal skills.
Number of participants needed: The exercise is designed for use with a class of up to 25 pupils or students, but can be used with lower numbers (min 5).
Time required: 5 school periods of roughly 35 - 40 minutes each or equivalent (including preliminary lesson).

Contents of package: teacher's guide and background booklet, introductory sheets, project sheets and work sheets.

Special facilities required: calculators, overhead projector, blank OHP transparencies and felt pens.

Outline of exercise: Power for Elaskay is a structured lesson on alternative energy built round a simulated case study. It is based on the hypothesis that Elaskay, an imaginary island supposedly located somewhere off the west coast of Scotland, has to replace its existing electricity generating plant over the next few years and has decided to meet its future electricity requirements by exploiting its natural energy resources — peat, solar energy, wind energy, tidal energy and hydro-electric power. The participants have to carry out technical and economic appraisals of each of these resources and then devise a 50-year rolling programme for meeting the island's future electricity needs.

POWER STATION PROJECT

Author: H I Ellington & E Addinall

Publisher: The Association for Science Education, College Lane, Hatfield, Herts AL10 9AA

First published: 1980

Type of exercise: manual simulated case study.

Courses for which exercise is suitable: The exercise was specially developed for use in the ASE's Science in Society AO level course, and is suitable for use in science in society and general studies courses at upper secondary and lower tertiary levels.

Educational aims: to make the participants aware of the basic technical principles and economics of electricity generation and of how power stations are planned; to cultivate interpretative, analytical, decision-making, communication and inter-personal skills.

Number of participants required: The exercise is designed for use with a class of up to 24 pupils or students, but can be used equally effectively with smaller numbers (min 6).

Time required: 3 school periods of roughly 35 - 40 minutes each or equivalent.

Contents of package: teacher's guide, student booklets, maps and other resource materials.

Special facilities required: calculators, overhead projector, blank OHP transparencies and felt pens.

Outline of exercise: The Power Station Project is a highly abridged, greatly simplified version of The Power Station Game (see page 144). Like the latter, it is based on the hypothesis that a policy decision has been reached to build a 2000 MW power station in a certain (imaginary) area. The object of the exercise is to draw up plans for three alternative schemes (a coal-fired station, an oil-fired station and a nuclear station) and then to decide which should be built.

PSTEP
Author: C Aust (available from the author at the School of Physics,
Robert Gordon's Institute of Technology, Aberdeen)
First published: 1978
Type of exercise: computer simulation.
Computer language: BASIC.
Courses for which exercise is available: physics, chemistry and engineering
courses at upper tertiary level.
Educational aims: to enable the student to investigate the phenomenon of
quantum-mechanical reflection and transmission of a beam of particles by
a potential step without the need for extensive numerical calculation.
Number of participants needed: 1 or 2.
Time required: ½ - 1 hour.
Contents of package: students' notes, computer program.
Special facilities required: interactive computer terminal, pocket calculator.
Outline of exercise: The program requests the height of the potential step
and the mass and energy of the incident particles. It responds by printing
the real part of the wave function for the incident, reflected and
transmitted beams, together with the coefficients of reflection and
transmission. The wave function may be displayed in tabular and/or
graphical form. The effect of varying each or all of the parameters can
be examined.

PWELL
Author: C Aust (available from the author at the School of Physics,
Robert Gordon's Institute of Technology, Aberdeen)
First published: 1978
Type of exercise: computer simulation.
Computer language: BASIC.
Courses for which exercise is suitable: physics, chemistry and engineering
courses at upper tertiary level.
Educational aims: to enable the student to investigate the dependence of
the energy eigenvalues available to a particle in a potential well of finite
depth on the depth and width of the well and the mass of the particle.
Number of participants needed: 1 or 2.
Time required: roughly 30 minutes.
Contents of package: students' notes, computer program.
Special facilities required: interactive computer terminal, pocket calculator.
Outline of exercise: The program requests the depth and width of the
potential well and the mass of the particle. It responds by printing the
quantum numbers and energy eigenvalues, measured from the top of the
well. The effect of varying each or all of the parameters can be examined.

RADIOACTIVE DECAY
Authors: B Samways & R D Masterton
Publisher: Schools Council, distributed by Edward Arnold (Publishers)
Ltd, London
First published: 1978

Price: part of a pack Computers in the Physics Curriculum, edited by
R D Masterton and D C F Chaundy, price £14.00
Type of exercise: computer simulation.
Computer language: BASIC.
Courses for which exercise is suitable: physics courses from middle
secondary to lower tertiary level.
Educational aims: to illustrate the exponential nature of radioactive decay;
to provide an introduction to the concept of a radioactive series and how
equilibrium in a series can occur; to teach the meaning of common
radioactive terminology.
Number of participants needed: The exercise can be used either
individually or with small groups.
Contents of package: teacher's notes, students' leaflets, computer program.
Special facilities required: interactive computer terminal or microcomputer.
Outline of exercise: This exercise combines laboratory investigations with
computerized models of simple and series radioactive decay phenomena.

SATELLITE MOTION
Inquiries: R Lewis, Educational Computing Section, Chelsea College,
Pulton Place, London SW6 5PR
First published: 1977
Type of exercise: computer simulation.
Computer language: BASIC or FORTRAN.
Courses for which exercise is suitable: physics courses at tertiary level.
Educational aims: to increase students' understanding of the relationship
between the energy of a satellite and the corresponding orbit.
Number of participants needed: The exercise can be used either
individually or with small groups.
Time required: 1 hour.
Contents of package: teacher's guide, students' notes, computer program.
Special facilities required: graphics terminal.
Outline of exercise: The student determines the launching angle and speed
of the satellite, although he may change the satellite speed, and thereby
the orbit, after the launch. The corresponding orbit is displayed on the
graphics terminal, with the corresponding energy diagram.

SCATTER (UNIT ON PARTICLE SCATTERING)
Author: J Harris
Publisher: Edward Arnold (Publishers) Ltd, London (Chelsea Science
Simulation Project)
First published: 1975
Price: £7.25
Type of exercise: computer simulation.
Computer language: BASIC.
Courses for which exercise is suitable: physics courses at upper secondary
and lower tertiary levels.
Educational aims: to introduce students to the notion that by bombarding
an object with particles one can learn something of its nature (shape,

size, etc) from the scattering it produces; to increase understanding of the use of models in physics; to give some appreciation of the achievement of Rutherford, Geiger and Marsden.

Number of participants needed: The exercise can be used either individually or with small groups.

Time required: 1½ - 2 hours.

Contents of package: teacher's guide, students' notes, computer program.

Special facilities required: interactive computer terminal or microcomputer.

Outline of exercise: This unit is designed to complement more traditional methods in teaching about the 'Rutherford' scattering experiment. There are three parts to the unit. In the first simulation 'marbles' are scattered by hard, massive objects of regular shape. The second involves particles scattered by a hard object or by an 'inverse-square scatterer'. The final simulation involves the scattering of alpha particles by a thin foil, using a simple nuclear model.

THE ELECTRIC CIRCUIT

Authors: R D Stamp & W Harrison

Publisher: Longman Group Ltd, Resources Unit, 9/11 The Shambles, York

First published: 1975

Price: £0.55

Type of exercise: board game.

Courses for which exercise is suitable: physics courses at middle to upper secondary school level.

Educational aims: to make participants aware of the roles played by electrical components (cells, ammeters, voltmeters, resistances, bulbs, etc) in electrical circuits; to help the participants gain experience of simple electrical circuit design.

Number of participants needed: 2 - 6 players.

Time required: 1 school double period of roughly 1¼ - 1½ hours.

Contents of package: A4 booklet, in which the outer 3 sheets (of thin card) form the board, playing materials and electrical circuit cards. The inner sheets (of paper) constitute an 8-page background reader/instructional booklet for the pupils.

Special facilities required: scissors, felt pens in 6 different colours, 2 matchsticks for mounting spinners.

Outline of exercise: The participants first select an individual electrical circuit card which they wish to study (each circuit is made up of 7 components). They then move round the board, which represents an electrical circuit containing a selection of common components connected in series and in parallel, collecting as they go the electrical components they need to build the circuit shown on their card. The players then calculate any meter shunt/multiplier values in their chosen circuit.

THE POWER STATION GAME

Authors: H I Ellington (editor), E Addinall, T Carnie, A G Garrow,
J Graham, K Jackson, N H Langton & J R Muckersie
Publisher: The Institution of Electrical Engineers, Savoy Place,
London WC2R 0BL
First published: 1976; revised 1979
Price: £25 (inc p & p)
Type of exercise: manual simulation game.
Courses for which exercise is suitable: physics, engineering, science, science
in society and general studies courses at upper secondary and lower
tertiary levels.
Educational aims: to make the participants aware of the technical
principles and economics of electricity generation and of how power
stations are planned; to cultivate interpretative, analytical, decision-making,
communication and inter-personal skills.
Number of participants needed: optimum number 18 (min 12; max 24).
Time required: roughly 2 - 2½ days.
Contents of package: Teacher's guide, introductory booklets, data
booklets, maps, instruction sheets, etc.
Special facilities required: calculators, overhead projector, blank OHP
transparencies and felt pens.
Outline of exercise: The Power Station Game is based on the hypothesis
that a policy decision has been reached to build a 2000 MW power station
in a certain (imaginary) area; the object of the exercise is to reach a
decision as to what type of station should be built (coal-fired, oil-fired
or nuclear) and where it should be sited. The participants are divided into
three competing teams, each of which has first to prepare and then
present a case for building one particular type of station. An independent
jury selects the best scheme, and the exercise is then brought to a
conclusion by holding a simulated public inquiry into this scheme.

THE SOLAR SYSTEM – SPACE PROBE

Authors: R D Stamp & W Harrison
Publisher: Longman Group Ltd, Resources Unit, 9/11 The Shambles,
York
First published: 1975
Price: £1.10
Type of exercise: board game.
Courses for which exercise is suitable: physics courses at upper secondary
level.
Educational aims: to help the participants understand the way in which
the relevant laws of physics can be applied to space travel within the
solar system.
Number of participants needed: 2 - 6 players.
Time required: 1 school double period of roughly 1¼ - 1½ hours.
Contents of package: 2 A4 booklets whose outer sheets (of thin card) are
used to prepare the board and other playing materials. The inner sheets
(of paper) constitute 2 - 8 page background readers/instructional booklets

for the pupils.

Special facilities required: scissors to prepare playing materials, matchsticks to mount spinners, electronic calculators.

Outline of exercise: The game is in two parts. In the first, the participants take the role of flight directors in charge of a manned, fact-finding probe to one of the planets. They have to calculate thrust requirements for their destination and prepare a reference trajectory which they hope to follow in the second part (they are given control sheets to fill in at this stage). The required trajectory is then plotted on their control sheet for future reference. In the second, they move over the board (which simulates the solar system) from the launch-point to the target planet and back to earth in stages, answering questions at each stage. The winner is the first probe to splash down, having successfully completed its mission.

Section 2: Chemistry-based exercises

A FERTILIZER TO ORDER (CHEMISTRY IN ACTION − 8)
Author: N Reid
Publisher: Heinemann Educational Books, London
First published: 1980
Price: part of a textbook *Chemistry About Us* by A H Johnstone, T I Morrison and N Reid.
Type of exercise: manual case study.
Courses for which exercise is suitable: chemistry courses at middle secondary level (specifically designed for Scottish O Grade).
Educational aims: to give the participants some basic knowledge of the chemical nature of fertilizers, awareness of some of the factors that influence choice of fertilizers, and awareness of the contribution of chemistry to food production; to help cultivate inter-personal, communication and data-handling skills.
Number of participants needed: groups of 3 or 4.
Time required: roughly 60 minutes.
Contents of package: part of textbook (one textbook required per participant).
Special facilities required: none.
Outline of exercise: Pupils consider data provided on a range of compounds which might be useful in designing a fertilizer. They have to select suitable compounds, and then carry out a costing in making a tonne of the fertilizer mixture.

A PROBLEM FOR THE CHEMIST
Author: N Reid
Publisher: Scottish Council for Educational Technology, Dowanhill, Victoria Crescent Road, Glasgow
First published: 1977
Price: £4.50
Type of exercise: manual case study.
Courses for which exercise is suitable: chemistry courses at middle

secondary level (designed specifically for Scottish O Grade).
Educational aims: to develop (a) an awareness of the importance of nitrogen and fertilizer production in world food problems; (b) an understanding of the key importance of research and its motivational aspects.
Number of partcipants needed: any number up to 20.
Time required: 1 school period of roughly 40 - 45 minutes.
Contents of package: tape and booklets.
Special facilities required: cassette player.
Outline of exercise: A tape guides the pupils through a short booklet, in which the history of nitrogen is discussed, and the present industrial practice outlined. The booklet ends with a series of questions that must be faced in the future.

BATCH OR FLOW
Author: N Reid
Publisher: Education Division, The Chemical Society, Burlington House, Piccadilly, London W1V 0BN
First published: 1979
Price: £1.50
Type of exercise: manual case study.
Courses for which exercise is suitable: chemistry courses at upper tertiary level.
Educational aims: by allowing students to take a series of simulated production and design decisions, to help them appreciate the way in which the chemical industry is radically different from the type of chemistry they encounter in traditional courses.
Number of participants needed: 6 or more.
Time required: roughly 3 hours.
Contents of package: teacher's guide, briefing and resource sheets for participants.
Special facilities required: none.
Outline of exercise: Students are faced with the problem of an explosive (PETN), which is obtained in an impure form from its synthesis. They have to select a solvent that is suitable for a possible purification process. The next step is to obtain pure crystals of PETN from this solvent. Finally, they have to consider two possible process routes (batch or flow), and make recommendations for research which should be considered.

CASE STUDIES IN CHEMICAL ENGINEERING
Publisher: ICI Educational Publications. Obtainable from The Kynoch Press, Thames House North, Millbank, London SW1P 4QG
First published: 1974
Price: £1 for 10 (or 25p per copy)
Type of exercise: manual case studies.
Courses for which exercise is suitable: chemistry courses at upper secondary level.
Educational aims: to help the participants appreciate the role of chemical

engineering in a modern technological society.

Number of participants needed: depends on number of copies available — no specific restriction.

Time required: up to 1 hour for each case study.

Contents of package: self-contained booklet.

Special facilities required: none.

Outline of exercise: The package contains 5 structured case studies which can be used as the basis of taught lessons or individual study. The case studies have the following titles:

1. Aniline: manufacture by continuous liquid phase hydrogenation.
2. Lime: the conversion of lime kilns from coal to gas firing.
3. New Protein: design and scale-up of a protein fermenter.
4. Tetrafluoroethylene: manufacture by steam pyrolysis.
5. Sulphuric Acid: removal of sulphuric acid mist from plant exhaust gases.

CHEM BINGO

Author: R L Gang

Publisher: Teaching Aids Co, 925 South 300 West, Salt Lake City, Utah 84101, USA

First published: 1971

Price: $6.00

Type of exercise: manual game.

Courses for which exercise is suitable: chemistry and general science courses at middle secondary level.

Educational aims: to help the participants memorize the symbols of different ions and elements.

Number of participants needed: The game can be used with a class of any size.

Time required: at the discretion of the teacher.

Contents of package: 4 pads of Bingo Game Sheets (consumable), 2 call lists (elements and ions)

Special facilities required: 25 cover pieces per player, or a pencil for each student.

Outline of exercise: The game is played in the same way as conventional Bingo, using the call lists included in the package.

CHEM CHEX

Author: R L Gang

Publisher: Teaching Aids Co, 925 South 300 West, Salt Lake City, Utah 84101, USA

First published: 1971

Price: $9.95

Type of exercise: a family of 5 board games.

Courses for which exercise is suitable: chemistry and general science courses at middle secondary level.

Educational aims: to develop an understanding of elements and ions and the way in which they combine to form chemical compounds.

Number of participants needed: 2 - 4.
Time required: 20 - 30 minutes.
Contents of package: checker (draughts) board, set of checkers (draughts), punch-out labels, chemical formulae chart, instruction booklet and periodic table.
Special facilities required: none.
Outline of exercise: Chem Chex is a family of five games that vary considerably in difficulty. They are all played with checker (draught)-like pieces on a checker (draught) board, using basically the same rules as checkers (draughts). The games end when all the pieces of one player or team have been captured, the winner being the player or team with *most points,* not most pieces.

CHEM CUBES ELEMENTS

Author: R L Gang
Publisher: Teaching Aids Co, 925 South 300 West, Salt Lake City, Utah 84101, USA
First published: 1972
Price: $7.00
Type of exercise: series of 8 manual games.
Courses for which exercise is suitable: chemistry and general science courses at middle secondary level.
Educational aims: to develop understanding of the elements and the use of the periodic table.
Number of participants needed: 4 - 6.
Time required: 30 - 45 minutes.
Contents of package: 6 wooden cubes, shaker cup, score pad, instruction booklet.
Special facilities required: periodic chart of elements (can be obtained free by writing to Publications Business Manager, Merck & Co Inc, Rahway, New Jersey 07065, USA).
Outline of exercise: Each person rolls the same cube, and determines the atomic number of the element rolled from the periodic table; the player with the highest score then plays first. Play consists of rolling cubes, and giving the answer for the combination rolled, points being scored for correct answers. One round is completed when everyone has had a turn to roll the cubes; 12 rounds complete the game. The 8 games in the series are arranged in a sequence covering the range of skills to be mastered. Students are also encouraged to devise their own games.

CHEMICAL CROSSWORDS

Authors: A G Hudson & S Hind
Publisher: Sigma Technical Press, 23 Dippons Mill Close, Tettenhall, Wolverhampton WV6 8HH
First published: 1977
Price: £1.25
Type of exercise: set of crossword puzzles.
Courses for which exercise is suitable: chemistry courses at middle and

upper secondary level (specifically designed for English O- and A-level courses).

Educational aims: to serve as a reinforcement and revision tool covering all aspects of chemistry included in the above courses.

Number of participants needed: can be used with groups of any size or for private study.

Time required: each crossword requires roughly 30 minutes to complete.

Contents of package: teacher's guide, set of 20 crosswords (1 copy of each) with solutions.

Special facilities required: none.

Outline of exercise: The participants complete the crosswords in the conventional way, all the answers being chemical formulae or numerical answers to chemical problems. The 20 puzzles in the pack are designed for use at different levels, and cover different branches of chemistry (organic and inorganic).

CHEMICAL ELEMENT GAME

Authors: K Shaw, P Hepburn & D Want

Publisher: Schools Council, distributed by Edward Arnold (Publishers) Ltd, London

First published: 1978

Price: part of a pack Computers in the Chemistry Curriculum, edited by K Shaw and D Want, price £15.00

Type of exercise: computer game.

Computer language: BASIC.

Courses for which exercise is suitable: chemistry courses at upper secondary and lower tertiary levels.

Educational aims: to encourage students to apply their knowledge of the properties of some elements and their compounds, and of trends in the periodic table.

Number of participants needed: The exercise can be used by competing individuals or competing small groups.

Contents of package: teacher's notes, students' leaflets, computer program.

Special facilities required: interactive computer terminal or microcomputer.

Outline of exercise: This is a competitive game which reviews the properties of some elements and their compounds. In the game students attempt to identify a mystery element from a total of 34. The aim is to identify each element using the minimum amount of information from the computer.

CHEMICAL FAMILIES

Authors: R D Stamp & W Harrison

Publisher: Longman Group Ltd, Resources Unit, 9/11 The Shambles, York

First published: 1975

Price: £0.55

Type of exercise: card (and board) game.

Courses for which exercise is suitable: chemistry courses at middle to upper secondary school level.

Educational aims: to help the participants remember and gain some understanding of the way in which the elements are grouped in families with similar physical and chemical properties in the periodic table.
Number of participants needed: 1 - 6 players.
Time required: 1 school period of roughly 35 - 40 minutes for each game.
Contents of package: A4 booklet, in which the two outer sheets (of thin card) are used to prepare the board and playing cards. The innermost sheets (of paper) constitute an 11-page background reader/instructional booklet for the pupils.
Special facilities required: scissors to prepare playing cards.
Outline of exercise: There are three different games in this package:
Game 1 — for 2 - 6 players. It is based on the card game known as sevens. The element cards are dealt (10 to each player) and each player first has to arrange his cards into periodic table order and then place them on the board (representing the periodic table) in his turn.
Game 2 — for 2 - 4 players. This is a rummy-type card game in which players have to collect groups of element cards and place them on the board. The winner is the first player to clear his hand (apart from 2 cards) or to complete group VIII.
Game 3 — for 1 player. A game of patience in which 3 rows of 9 cards are dealt (and laid on top of each other) face downwards. A fourth row of 9 cards is then placed on top, this time face upwards. The player then attempts to complete all groups of the peridoic table in descending order.

CHEMINOES
Publisher: Chemical Teaching Aids, Letham, Ladybank, Fife, Scotland KY7 7RN
Price: £0.55
Type of exercise: card game.
Courses for which exercise is suitable: chemistry and general science courses at lower to middle secondary level.
Educational aims: to make the participants aware of the role of valency in the combining of elements and radicals to make compounds.
Number of participants required: 2 - 6 players.
Time required: separate sessions of 30 minutes, repeated as required.
Contents of package: set of domino-type cards.
Special facilities required: none.
Outline of exercise: Different elements and/or radicals are identified on each card together with their valency requirements. The pupils have to match the ends of the cards in domino fashion, using valencies instead of spots.

CHEMISTRY CASE STUDIES
Publisher: Joint Matriculation Board, Manchester M15 6EU
Price: £1.50 per copy
Type of exercise: manual case studies.

Courses for which exercise is suitable: chemistry courses at upper secondary level.

Educational aims: to help the participants develop an appreciation of the social and economic aspects of chemical technology.

Number of participants needed: depends on number of copies available — no specific restriction.

Time required: up to 4 hours for each case study.

Contents of package: self-contained booklet.

Special facilities required: none.

Outline of exercise: The package contains 6 structured case studies which the participants are expected to work through in their own time, class teaching being devoted to discussion of principles and conclusions. The case studies have the following titles:

1. A problem of effluent disposal
2. The use and abuse of drugs
3. Five routes to phenol
4. Sulphur and the environment
5. The spray-steel process
6. Salt-based industries.

CHEMISTRY IN OUR LIVES

Author: N Reid

Publisher: Scottish Council for Educational Technology, Dowanhill, Victoria Crescent Road, Glasgow

First published: 1977

Price: £1.25

Type of exercise: competitive lecturette presentation (a game).

Courses for which exercise is suitable: chemistry courses at middle secondary level (designed specifically for Scottish O Grade).

Educational aims: to provide (a) an awareness of the impact of chemistry on five main areas of daily life; (b) experience in presenting a reasoned and compelling argument; (c) experience in judging the case put forward by others.

Number of participants needed: 10 - 20 players.

Time required: 1 school double period of roughly 1¼ - 1½ hours.

Contents of package: paperwork.

Special facilities required: none.

Outline of exercise: 5 groups are formed, each group being given information on a different topic. Each group has to prepare a short lecturette, and 'Eurovision voting' follows the presentation.

CHEMSYN

Authors: G Eglinton & J R Maxwell

Publisher: Heyden & Son Ltd, Spectrum House, Hillview Gardens, London NW4 2JQ

First published: 1972

Price: £1.80

Type of exercise: card game.

Courses for which exercise is suitable: courses in organic chemistry at upper secondary and lower tertiary levels.

Educational aims: to help reinforce and consolidate basic ideas regarding (i) the names and classes of simple organic compounds; (ii) basic functional group chemistry; (iii) the basic principles of stereochemistry, and (iv) simple synthetic chemistry.

Number of participants needed: The game can be played by a single person or by a group of up to 5; it can be used with a class of any size by employing more than one pack.

Time required: The various games in which the Chemsyn pack can be used have a minimum playing time of the order of 20 minutes.

Contents of package: pack of 50 cards, game manual (for use by teacher or students).

Special facilities required: none.

Outline of exercise: Chemsyn consists of a pack of 50 double-sided cards, each depicting a representative member of a particular class of organic compounds and giving basic information about its principal reactions. The object of the game is to use the cards to build up a sequence showing how the various compounds (or the classes to which they belong) can be inter-converted. In solo Chemsyn, a single player works systematically through the pack in solitaire fashion, trying to add each card to the overall sequence. In group Chemsyn, the players operate in rummy fashion, each trying to get rid of the cards in his hand by adding them to a common sequence.

CHEM TRAK

Author: R L Gang

Publisher: Teaching Aids Co, 925 South 300 West, Salt Lake City, Utah 84101, USA

First published: 1975

Price: $1.95

Type of exercise: board game/case study for individuals or for a whole class.

Courses for which exercise is suitable: chemistry and general science courses at middle secondary level.

Educational aims: to help develop understanding of the way in which elements and ions combine to form compounds.

Number of participants needed: 1 - 30, depending on the way in which the material is being used.

Time required: at the discretion of the teacher.

Contents of package: 1 Chem-Trak board (1 needed per pupil).

Special facilities required: none.

Outline of exercise: The Chem-Trak board can be used either as an individualized tutor or as the basis of a competitive exercise. It reviews 263 different chemical formulae and the names of 552 different compounds, and can be used to do a variety of different things, ie

    (i)   to name ions and elements around the track

    (ii)   to name compounds formed using ions and elements

(iii)   to write correct formulae and names of compounds
(iv)   to write names of acids
 (v)   to study ionic and combination reaction equations
(vi)   to study single replacement reactions.

CHOOSE A FIBRE (CHEMISTRY IN ACTION – 10)

Author: N Reid
Publisher: Heinemann Educational Books, London
First published: 1980
Price: part of a textbook *Chemistry About Us* by A H Johnstone,
T I Morrison and N Reid.
Type of exercise: manual case study.
Courses for which exercise is suitable: chemistry courses at middle
secondary level (specifically designed for Scottish O Grade).
Educational aims: to make the participants aware of the relationships
between property, use and personal judgement in relation to fibres, and
help them appreciate that real-life decisions are often based on
compromise; to help cultivate inter-personal, communication and
data-handling skills.
Number of participants needed: groups of 3 or 4.
Time required: roughly 75 minutes.
Contents of package: part of textbook (one textbook required per
participant).
Special facilities required: none.
Outline of exercise: Pupils are given data on a range of unnamed fibres.
They discuss the behaviour required for a fibre for a series of uses, and
then attempt to select the best fibre(s) from the given range.

CHOOSING AN ANAESTHETIC

Author: N Reid
Publisher: Scottish Council for Educational Technology, Dowanhill,
Victoria Crescent Road, Glasgow
First published: 1979
Price: £2.25
Type of exercise: simulation.
Courses for which exercise is suitable: chemistry courses at middle
secondary level (designed specifically for Scottish O Grade).
Educational aims: to provide (a) understanding of the basic criteria in
selecting anaesthetics; (b) experience in sifting data; (c) confidence in
decision making; (d) awareness of the contribution of chemistry to
medical advance.
Number of participants needed: up to 20, in groups.
Time required: 1 school double period of roughly 1¼ - 1½ hours.
Contents of package: paperwork.
Special facilities required: none.
Outline of exercise: Working in groups, pupils are taken through a series of
decisions to reach new compounds for use as anaesthetics. In a similar way,
they relive the great triumph as halothane was sought and discovered.

COMPETITION AMONG THE METALS

Authors: R D Stamp and W Harrison
Publisher: Longman Group Ltd, Resources Unit, 9/11 The Shambles, York
First published: 1975
Price: £0.55
Type of exercise: board game.
Courses for which exercise is suitable: chemistry courses at middle to upper secondary level.
Educational aims: to help the participants understand the differing reactivity of metals and to demonstrate the reactivity series.
Number of participants needed: 2, 4 or 6 players.
Time required: 1 school period of roughly 35 - 40 minutes.
Contents of package: A4 booklet whose outer sheets (of thin card) are used to prepare the board and other playing materials. The inner sheets (of paper) constitute an 8-page background reader/instructional booklet for the pupils.
Special facilities required: scissors to prepare playing materials.
Outline of exercise: The board represents a court on which the players (metals) play a badminton-type game. Whether or not a service is called good depends upon the state of occupancy of the receiving square in the opponents' court. The idea is to try and score points, with metals high in the reactivity series having precedence over those below them. The rules for receiving and/or returning service are similar. Play continues until one team scores 9 points and is 2 points clear.

COMPOUNDS

Author: P Ploutz
Publisher: Union Printing Co Inc, 17 W Washington Street, Athens, Ohio 45701, USA
First published: 1974
Price: $8.00 (reductions available for orders of 10 or more)
Type of exercise: card game.
Courses for which exercise is suitable: chemistry and general science courses at lower and middle secondary level.
Educational aims: to reinforce work on the formation of chemical compounds (based on 21 of the commonest elements).
Number of participants needed: 2 - 6
Time required: This varies according to the number and age of the participants, but is about 20 minutes.
Contents of package: 80 element cards, 4 jokers, instruction booklet.
Special facilities required: none.
Outline of exercise: Compounds is designed to reinforce the participants' knowledge of chemical compounds and their formation. The element cards are colour-coded to indicate metals, non-metals, gases, etc and, as players become more skilled, they are able to relate cards to the natural physical state of the element as well as to its location in the periodic table. Compounds complements Elements, another game produced by the same company, although each game is completely self-contained.

CONTACT WITH CLEANER AIR (CHEMISTRY IN ACTION – 7)

Author: N Reid
Publisher: Heinemann Educational Books, London
First published: 1980
Price: part of a textbook *Chemistry About Us* by A H Johnstone,
T I Morrison and N Reid.
Type of exercise: manual case study
Courses for which exercise is suitable: chemistry courses at middle
secondary level (specifically designed for Scottish O Grade).
Educational aims: to make the participants aware of the complexity of
an apparently simple industrial process, of the need for pollution
legislation, and of the impact of such legislation on industry; to help
cultivate inter-personal, communication, problem-solving and decision-
making skills.
Number of participants needed: groups of 3 or 4.
Time required: roughly 50 minutes.
Contents of package: part of textbook (one textbook required for each
participant).
Special facilities required: none.
Outline of exercise: The Contact Process is often presented as a very
simple process. In this exercise, pupils are exposed to the real-life
industrial problems associated with the process, and they are led to a
series of decisions in designing a factory that is economic and able to
meet current pollution legislation.

DIS-SOLVING A PROBLEM (CHEMISTRY IN ACTION – 11)

Author: N Reid
Publisher: Heinemann Educational Books, London
First published: 1980
Price: part of a textbook *Chemistry About Us* by A H Johnstone,
T I Morrison and N Reid.
Type of exercise: manual case study.
Courses for which exercise is suitable: chemistry courses at middle
secondary level (specifically designed for Scottish O Grade).
Educational aims: to give the participants some insight into the sort of
problems encountered in the chemical industry; to help cultivate data-
handling, communication, problem-solving and decision-making skills.
Number of participants needed: groups of 3 or 4.
Time required: roughly 80 minutes.
Contents of package: part of textbook (one textbook required for each
participant).
Special facilities required: none.
Outline of exercise: This exercise considers the industrial purification of
PETN (a common explosive). Pupils have to select the best solvent for the
purification process, and then seek to use the solvent in such a way that
the purification process is efficient and economic. The problem is a
difficult one, in which pupils have to think beyond their limited school
laboratory experiences.

ELECTROCHEMICAL CELLS

Authors: K Shaw, P Hepburn, D Spiers and M Poore
Publisher: Schools Council, distributed by Edward Arnold (Publishers)
Ltd, London
First published: 1978
Price: part of a pack Computers in the Chemistry Curriculum, edited by
K Shaw and D Want, price £15.00
Type of exercise: computer simulation.
Computer language: BASIC.
Courses for which exercise is suitable: chemistry courses at upper
secondary and lower tertiary levels.
Educational aims: to provide students with an introduction to some of
the important features of electrochemical cells.
Number of participants needed: The exercise can be used either
individually or with small groups.
Contents of package: teacher's notes, students' notes, computer program.
Special facilities required: interactive computer terminal or microcomputer.
Outline of exercise: This exercise simulates experiments which investigate
factors that influence electrode system potentials, including redox cells
and pH sensitive reactions.

ELEMENT CARDS

Author: J I Lipson
Publisher: Heyden & Son Ltd, Spectrum House, Hillview Gardens,
London NW4 2JQ
First published: 1972
Price: £2.60
Type of exercise: card games.
Courses for which exercise is suitable: chemistry courses at lower
secondary to tertiary level.
Educational aims: to help participants appreciate the importance of the
periodic classification of the elements; to cultivate interpretative and
analytical skills.
Number of participants needed: 1 - 10 (depending on game chosen).
Time required: sessions of roughly 30 minutes.
Contents of package: booklet, pack of 104 cards.
Special facilities required: none.
Outline of exercise: The package consists of a series of 9 different card
games varying in difficulty and sophistication, including rummy, cribbage,
solitaire, pairs, etc. Each card contains detailed information about one
element, ie atomic number, name, valency states, melting and boiling
points, chemical symbol, atomic weight, isotopes and electronic structure.
Participants have to use some or all of this information to play
effectively in each of the games.

ELEMENTS
Author: P Ploutz
Publisher: Union Printing Co, Inc, 17 W Washington Street, Athens, Ohio 45701, USA
First published: 1970
Price: $14.00 (reductions available for orders of 10 or more)
Type of exercise: board game.
Courses for which exercise is suitable: chemistry and general science courses at lower and middle secondary level.
Educational aims: to help the participants become familiar with the names and properties of the elements and their positions in the periodic table.
Number of participants needed: 2 - 6.
Time required: This varies according to the particular game being played, the number of participants and the age of the participants, but is normally about 30 minutes.
Contents of package: board, 105 element chips, instruction book.
Special facilities required: none.
Outline of exercise: Elements is in fact a family of 4 games and 8 variations, all using the same playing materials. The various games are suitable for use at a variety of levels, and with students of a wide range of abilities. The basic principle underlying most of the games involves placing the element chips in their correct positions in the periodic table. Elements complements Compounds, produced by the same company.

ELEMENTS IN THE EARTH
Author: N Reid
Publisher: Scottish Council for Educational Technology, Dowanhill, Victoria Crescent Road, Glasgow
First published: 1979
Price: £5.25
Type of exercise: manual case study.
Courses for which exercise is suitable: chemistry courses at middle secondary level (designed specifically for Scottish O Grade).
Educational aims: to provide (a) a more detailed working knowledge of the periodic table; (b) an appreciation of the periodic table as a useful classifying tool; (c) an appreciation of patterns in the discovery of the elements and their sources.
Number of participants needed: up to 20, in small groups.
Time required: 1 double school period of roughly 1¼ - 1½ hours.
Contents of package: tape, OHP overlays, paperwork.
Special facilities required: cassette player, OHP.
Outline of exercise: Pupils start by working in groups to collate data on the dates of discovery, the main sources and the percentage occurrence of the elements. They then relate this data to the periodic table, looking for patterns and trying to interpret these patterns.

ENERGY FOR THE FUTURE

Author: N Reid

Publisher: Scottish Council for Educational Technology, Dowanhill, Victoria Crescent Road, Glasgow

First published: 1977

Price: £5.00

Type of exercise: manual exercise — programmed learning format.

Courses for which exercise is suitable: chemistry and general science courses at middle secondary level.

Educational aims: to provide (a) an insight into the kind of problems facing a future society; (b) an appreciation of the interplay of resources, technology and economic viability.

Number of participants needed: 1 - 20.

Time required: 1 school period of roughly 40 minutes.

Contents of package: booklet.

Special facilities required: none.

Outline of exercise: An exercise, set in programmed learning format, which considers the problem in the year 2000, when gas is running out. Is there a chemical contribution to the answer to the problem?

ENERGY — PAST, PRESENT AND FUTURE (CHEMISTRY IN ACTION — 3)

Author: N Reid

Publisher: Heinemann Educational Books, London

First published: 1980

Price: part of a textbook *Chemistry About Us* by A H Johnstone, T I Morrison and N Reid.

Type of exercise: manual case study.

Courses for which exercise is suitable: chemistry courses at middle secondary level (specifically designed for Scottish O Grade).

Educational aims: to make the participants aware of past and present primary sources of energy in Britain, of the increasing demand for energy, and of possible future trends; to cultivate interpretative and decision-making skills.

Number of participants required: groups of 3 or 4.

Time required: roughly 30 minutes.

Contents of package: part of textbook (one textbook required for each participant).

Special facilities required: none.

Outline of exercise: Pupils are given data about the energy use pattern in the UK for the past two decades. Working in groups, they first learn how to interpret cumulative graphs, and then attempt to extend the graph 30 years into the future. It is essentially a simple long-range forecasting exercise.

FOCUS ON LEAD

Author: N Reid

Publisher: Scottish Council for Educational Technology, Dowanhill, Victoria Crescent Road, Glasgow

First published: 1977
Price: £4.50
Type of exercise: manual case study.
Courses for which exercise is suitable: chemistry courses at middle
secondary level (designed specifically for Scottish O Grade).
Educational aims: to develop (a) a knowledge of the main facts about the
extraction, purification, and uses for lead; (b) an awareness of the social
contribution of lead and environmental costs; (c) an appreciation that
removal of lead from the environment is expensive.
Number of participants needed: any number up to 20.
Time required: 1 double school period of roughly 1¼ - 1½ hours.
Contents of package: tape and booklets.
Special facilities required: cassette player.
Outline of exercise: A tape guides pupils through a short booklet which
discusses the chemical and social importance of lead. The tape asks
questions which the group have to discuss. The exercise concludes with
short essays.

FORMULON
Publisher: Chemistry Teaching Aids, Letham, Ladybank, Fife KY7 7RN
First published: 1973
Price: £1.95
Type of exercise: card game.
Courses for which exercise is suitable: courses in basic inorganic chemistry
at lower or middle secondary level.
Educational aims: to reinforce and consolidate basic ideas regarding the
different types of atoms and ions and the ways in which they combine to
form ionic and covalent chemical compounds.
Number of participants needed: Each pack can be used with a group of
up to 8 players and, by using more than 1 pack, the exercise can be used
with a class of any size.
Time required: Each round of the game takes roughly 5 - 10 minutes, and
any number of rounds may be played in a given session.
Contents of package: pack of 100 cards, instruction leaflet (for use by
teacher or students).
Special facilities required: none.
Outline of exercise: The Formulon pack consists of 100 cards, each
representing a particular type of atom or ion, a multiplier (2 or 3) or a
joker. The players are dealt hands of 10 cards and, playing in rummy
fashion, have to try to get rid of their cards by forming valid chemical
compounds. The winner of each round is the first player to get rid of all
his cards, the other players collecting penalty points according to the
number and type of cards they have left. The overall winner is the player

with the smallest number of penalty points after a number of rounds have been completed.

FROM VISIBLE TO INVISIBLE (CHEMISTRY IN ACTION – 6)
Author: N Reid
Publisher: Heinemann Educational Books, London
First published: 1980
Price: part of a textbook *Chemistry About Us* by A H Johnstone,
T I Morrison and N Reid.
Type of exercise: manual case study.
Courses for which exercise is suitable: chemistry courses at middle secondary level (specifically designed for Scottish O Grade).
Educational aims: to help the participants appreciate how currently accepted ideas developed; to help develop problem-solving, communication and decision-making skills.
Number of participants needed: groups of 3 or 4.
Time required: roughly 40 minutes.
Contents of package: part of textbook (one textbook required per person).
Special facilities required: none.
Outline of exercise: Pupils are led in a simple way through the problems that faced Gay-Lussac as he studied gas reactions. Working as a group, they have to identify his fundamental problem and then, imagining that they could go back in time and talk to him, try to explain where his ideas were incorrect.

GAS CHROMATOGRAPHY
Author: D Want
Publisher: Schools Council, distributed by Edward Arnold (Publishers) Ltd, London
First published: 1978
Price: part of a pack Computers in the Chemistry Curriculum, edited by K Shaw and D Want, price £15.00
Type of exercise: Computer simulation.
Computer language: BASIC.
Courses for which exercise is suitable: chemistry courses at upper secondary and lower tertiary levels.
Educational aims: to allow students to 'experiment' with gas chromatography in a way which would preclude the use of sensitive and expensive equipment.
Number of participants needed: The exercise can be used either individually or with small groups.
Contents of package: teacher's notes, students' notes, computer program.
Special facilities required: interactive computer terminal or microcomputer.
Outline of exercise: This exercise involves simulations of the separation of some organic compounds to show the application of gas chromatography to qualitative and quantitative analysis.

HABER (UNIT ON AMMONIA SYNTHESIS)

Authors: R Edens & K Shaw
Publisher: Edward Arnold (Publishers) Ltd, London (Chelsea Science Simulation Project).
First published: 1978
Price: £7.25
Type of exercise: computer simulation.
Computer language: BASIC.
Courses for which exercise is suitable: chemistry courses at middle and upper secondary level.
Educational aims: to enable students to discover how various conditions (temperature, pressure, catalyst and reactant concentration ratios) influence the course of the Haber process reaction; to develop good practice in the design of investigations.
Number of participants needed: The exercise can be used either individually or with small groups.
Time required: 1½ - 2 hours.
Contents of package: teacher's guide, students' notes, computer program.
Special facilities required: computer terminal or microcomputer.
Outline of exercise: The simulation is divided into two parts. In the first part, students investigate the effect of varying temperature and pressure and the molar ratio of hydrogen to nitrogen on the percentage yield of ammonia. In the second investigation, students look at the effect of varying temperature, pressure and catalyst on the rates of both the forward and reverse reactions, ie on the rate of attaining equilibrium.

HOMOGENEOUS EQUILIBRIUM

Authors: K Shaw & R Edens
Publisher: Schools Council, distributed by Edward Arnold (Publishers) Ltd, London.
First published: 1978
Price: part of a pack Computers in the Chemistry Curriculum edited by K Shaw and D Want, price £15.00
Type of exercise: computer simulation.
Computer language: BASIC.
Courses for which exercise is suitable: chemistry courses at upper secondary and lower tertiary levels.
Educational aims: to establish the concept of dynamic equilibrium, to develop an understanding of the relationship between the stoichiometric equation for a reaction and the expression for the equilibrium constant; to help students gain experience in the design of experiments.
Number of participants needed: The exercise can be used either individually or with small groups.
Contents of package: teacher's notes, students' leaflets, computer program.
Special facilities required: interactive computer terminal or microcomputer.
Outline of exercise: The exercise involves an investigation of acid-alcohol

equilibrium and some gas-phase equilibria leading to the equilibrium law in terms of concentration.

IONICS

Publisher: Science Systems Ltd, 173 Southampton Way, London SE5 7EJ
First published: 1972
Price: £0.75
Type of exercise: card game.
Courses for which exercise is suitable: chemistry courses at lower to middle secondary level.
Educational aims: to make the participants aware of the ways in which compounds can be formed using ionic bonding.
Number of participants needed: 2 - 5.
Time required: 1 school period of roughly 40 minutes.
Contents of package: leaflet, set of 52 cards.
Special facilities required: none.
Outline of exercise: A series of three rummy-type card games in which the participants make up ionic compounds. Variations such as naming the compounds made and the facility of neutralization and double decomposition can be introduced at the higher levels.

ISOTOPES

Author: N Reid
Publisher: Scottish Council for Educational Technology, Dowanhill, Victoria Crescent Road, Glasgow
First published: 1977
Price: £7.00
Type of exercise: manual case study.
Courses for which exercise is suitable: chemistry courses at middle secondary level (designed specifically for Scottish O Grade).
Educational aims: to develop (a) communication skills within a peer group; (b) a willingness to argue on the basis of provided evidence; and to provide (a) an historic perspective on the nuclear industry; (b) the beginnings of an awareness of the complexities of nuclear decision making.
Number of participants needed: up to 24, in 4 groups.
Time required: 1 school double period of roughly 1¼ - 1½ hours.
Contents of package: tapes, slides, paperwork.
Special facilities required: cassette player, slide projector.
Outline of exercise: Pupils are introduced to the topic of radioisotopes through a short tape/slide sequence. They then work in groups of 6 to examine the issues surrounding the production and uses of radioisotopes today.

ISOTOPES IN OUR LIVES

Author: N Reid
Publisher: Scottish Council for Educational Technology, Dowanhill, Victoria Crescent Road, Glasgow
First published: 1977

Price: £1.50
Type of exercise: manual case study.
Courses for which exercise is suitable: chemistry courses at upper secondary level (designed specifically for Scottish O Grade).
Educational aims: to provide an understanding of the basic facts about current nuclear technology; to develop (a) communication skills within a peer group; (b) an awareness of the complexities of decision taking within the nuclear industry.
Number of participants needed: up to 24, in 4 groups.
Time required: 1 school double period of roughly 1¼ - 1½ hours.
Contents of package: paperwork.
Special facilities required: none.
Outline of exercise: Pupils work in groups of 6 to examine the issues surrounding the production and uses of radioisotopes.

LATTICE ENERGY

Authors: R Edens & K Shaw
Publisher: Schools Council, distributed by Edward Arnold (Publishers) Ltd, London
First published: 1978
Price: part of a pack Computers in the Chemistry Curriculum edited by K Shaw and D Want, price £15.00
Type of exercise: computer simulation.
Computer language: BASIC.
Courses for which exercise is suitable: chemistry courses at upper secondary and lower tertiary levels.
Educational aims: to develop an understanding of (a) the concept of lattice energy and the factors which affect it; and (b) the use and limitations of the Born-Mayer equation as a mathematical model.
Number of participants needed: The exercise can be used either individually or with small groups.
Contents of package: teacher's notes, students' leaflets, computer program.
Special facilities required: interactive computer terminal or microcomputer.
Outline of exercise: Ionic crystal structures are examined with regard to the charge on the ions, their size and their co-ordination. The limitations of the simple model in some cases are also investigated.

LINK-UP

Authors: R D Stamp & W Harrison
Publisher: Longman Group Ltd, Resources Unit, 9/11 The Shambles, York
First published: 1975
Price: £0.55
Type of exercise: manual exercise.
Courses for which exercise is suitable: chemistry courses at middle to upper secondary level.
Educational aims: to help the participants understand how atoms bond together to form compounds by getting them to (a) build up selected

electronic structures; (b) build compounds from selected atoms; and (c) demonstrate the bonding arrangement of electrons in selected compounds.

Number of participants needed: 4 players working in pairs.

Time required: 3 school periods of roughly 35 - 40 minutes each.

Contents of package: A4 booklet whose outer sheets (of thin card) are used to prepare the playing materials. The inner sheets (of paper) constitute a 12-page background reader/instructional booklet for the pupils.

Special facilities required: scissors to prepare playing materials, 2 matchsticks for mounting spinners, waste discs from a file-punch (preferably in two colours) to represent electrons.

Outline of exercise: There are 3 stages in this exercise. In the first, the pupils have to collect 3 nuclear charges (of different values) and the appropriate numbers of electrons to build up 3 different atomic structures, by using the 2 spinners and acting on the instructions indicated. In the second, the pupils, now working in pairs, use the atomic structures they obtained in stage 1, to prepare a list of as many known compounds (based on their atoms) as they can. In the final stage, again working in pairs, the pupils have to demonstrate the type of bonding in 3 of the compounds from their list produced in stage 2.

MAGNESIUM FROM THE SEA

Author: N Reid

Publisher: Scottish Council for Educational Technology, Dowanhill, Victoria Crescent Road, Glasgow

First published: 1979

Price: £4.50

Type of exercise: manual case study.

Courses for which exercise is suitable: chemistry courses at middle secondary level (designed specifically for Scottish O Grade).

Educational aims: to provide (a) general knowledge of the extraction process to obtain magnesium from the sea; (b) experience of data-handling group discussion and decision making; (c) an awareness of the complexities of industrial processes.

Number of participants needed: up to 20, in small groups.

Time required: 1 school double period of roughly 1¼ - 1½ hours.

Contents of package: tape, booklets, paperwork.

Special facilities required: cassette player.

Outline of exercise: Pupils are introduced to the possibility of extracting magnesium from the sea in the first few pages of the booklet, the tape acting as a guide. They then work in groups to take a series of decisions in planning a possible extraction process to obtain magnesium oxide. The second section of the booklet completes the exercise by providing more information on current technology.

METALS AT WORK (CHEMISTRY IN ACTION – 4)
Author: N Reid
Publisher: Heinemann Educational Books, London
First published: 1980
Price: part of a textbook *Chemistry About Us* by A H Johnstone,
T I Morrison and N Reid.
Type of exercise: manual case study.
Courses for which exercise is suitable: chemistry courses at middle
secondary level (specifically designed for Scottish O Grade).
Educational aims: to help the participants appreciate the range of
properties associated with metals and relate properties to usage; to help
develop interpretative and communication skills.
Number of participants required: groups of 3 or 4.
Time required: roughly 40 minutes.
Contents of package: part of textbook (1 textbook required for each
participant).
Special facilities required: none.
Outline of exercise: Given data on a range of metals, pupils have to decide
the behaviour characteristics that are required for various uses, and then
compile a short-list of suitable metals.

MOLE BINGO
Author: A Armstrong
Publisher: Sigma Technical Press, 23 Dippons Mill Close, Tettenhall,
Wolverhampton WV6 8HH
First published: 1972
Price: £1.25
Type of exercise: set of manual games.
Courses for which exercise is suitable: chemistry and general science
courses at lower and middle secondary level.
Educational aims: to give practice in answering questions relating to
calculations on molar quantities.
Number of participants needed: 2 - 5.
Time required: roughly 30 minutes.
Contents of package: teacher's guide and resource material, sets of
problem cards.
Special facilities required: OHP (optional).
Outline of exercise: Mole Bingo is a set of 4 games graded in order of
increasing difficulty. All 4 games are based on conventional Bingo, the
participants having to answer questions on simple molar calculations
in order to cross off the numbers on their matrices, the winner being
the first to complete a line.

MOLE POKER
Author: A Armstrong
Publisher: Sigma Technical Press, 23 Dippons Mill Close, Tettenhall,
Wolverhampton WV6 8HH
First published: 1978

Price: £1.25

Type of exercise: set of card games.

Courses for which exercise is suitable: chemistry and general science courses at middle secondary level.

Educational aims: to give practice in calculations relating to molar quantities and chemical formulae.

Number of participants needed: 1 - 4.

Time required: roughly 30 minutes.

Contents of package: teacher's guide and resource material, cardboard cut-outs for preparing poker-type dice.

Special facilities required: none.

Outline of exercise: Mole Poker is similar to Mole Bingo, but is slightly more advanced. It consists of a series of games, played with special dice, in which the participants have to be able to answer questions relating to molar quantities and chemical formulae in order to play effectively.

PARTICLE IN A POTENTIAL WELL

Inquiries: R Lewis, Educational Computing Section, Chelsea College, Pulton Place, London SW6 5PR

First published: 1977

Type of exercise: computer simulation.

Computer language: BASIC or FORTRAN.

Courses for which exercise is suitable: chemistry and physics courses at tertiary level.

Educational aims: to give students a non-mathematical method for good conceptual understanding of the Schrödinger equation.

Number of participants needed: The exercise can be used either individually or with small groups.

Time required: 1 hour.

Contents of package: teacher's guide, students' notes, computer program.

Special facilities required: graphics terminal.

Outline of exercise: The student specifies the particle mass, well width and depth. The energy levels are displayed for the particle in the chosen square well and also in a square well of infinite depth.

POINT FIELDS

Authors: H I Ellington, A G Garrow & J R Muckersie

Publisher: The Institution of Electrical Engineers, Savoy Place, London WC2R 0BL

First published: 1980

Price: £9.00 (inc p & p)

Type of exercise: manual simulation game.

Courses for which exercise is suitable: chemistry, science in society and general studies courses at upper secondary and lower tertiary levels.

Educational aims: to make the participants aware of the complexity and importance to society of a large petrochemical plant and of the social role and procedure of a public inquiry; to cultivate communication and debating skills.

Number of participants needed: The exercise is designed for use with a class of up to 32 pupils or students, but can be used with smaller numbers (min 12).

Time required: 1½ - 3 hours, depending on the numbers involved and on the level and ability of the participants.

Contents of package: teacher's guide, role sheets and resource material for participants.

Special facilities required: overhead projector, blank OHP transparencies and felt pens.

Outline of exercise: Point Fields is a role-playing simulation exercise. It is based on the hypothesis that United Petrochemicals Ltd (a subsidiary of an imaginary major oil company) plan to build a large petrochemical plant at Point Fields, a coastal site somewhere in the east of Scotland, and that a public inquiry into the proposed scheme has been called. The exercise takes the form of this public inquiry. The participants are allocated roles supporting or objecting to the planned development, the various arguments being assessed by an independent 'reporter' who then decides whether or not the scheme should go ahead.

POLAR OR IONIC? (CHEMISTRY IN ACTION – 5)

Author: N Reid

Publisher: Heinemann Educational Books, London

First published: 1980

Price: part of a textbook *Chemistry About Us* by A H Johnstone, T I Morrison and N Reid.

Type of exercise: manual case study.

Courses for which exercise is suitable: chemistry courses at middle secondary level (specifically designed for Scottish O Grade).

Educational aims: to help the participants develop a critical approach to answering questions; to help them appreciate the role of experiments in scientific methodology and what constitutes a good experiment.

Number of participants required: either small groups or a class of pupils working with the teacher.

Time required: roughly 50 minutes.

Contents of package: part of textbook (1 textbook required for each participant).

Special facilities required: none.

Outline of exercise: The package looks at the bonding in aluminium chloride, using a conversational presentation. The main purpose is to allow pupils to appreciate the place of experimentation in science, and pupils have to take a series of decisions to this end.

POLYWATER

Authors: F Percival & A H Johnstone

Publisher: Education Division, The Chemical Society, Burlington House, Piccadilly, London W1V 0BN

First published: 1978

Price: £2.50

Type of exercise: manual case study.

Courses for which exercise is suitable: chemistry courses at upper tertiary level.

Educational aims: to develop library and communication skills by studying an actual example of chemical research; to help develop a healthy scepticism towards the written word.

Number of participants needed: groups of 6 - 8 people.

Time required: 3 hours (minimum)

Contents of package: teacher's guide, resource material for participants (A4 booklets)

Special facilities required: access to chemistry library of university standard.

Outline of exercise: Polywater is the case history of the polywater phenomenon, and takes the form of a structured library study/discussion exercise. The participants look up, study and precis papers from the early, middle and late phases of work on the phenomenon, discussing their findings with their colleagues at each stage of the work.

PROPERTIES AND SUBSTANCES

Author: A Armstrong

Publisher: Chemical Teaching Aids, Letham, Ladybank, Fife, Scotland KY7 7RN

First published: 1974

Price: £1.50

Type of exercise: card game.

Courses for which exercise is suitable: chemistry and general science courses at middle secondary level.

Educational aims: to make the participants aware of the basic properties of a range of elements and compounds.

Number of participants needed: 2 - 8 players.

Time required: 1 school period of roughly 40 minutes.

Contents of package: leaflet, set of cards.

Special facilities required: chemical data book for use as reference.

Outline of exercise: Properties and Substances is a card game which is a cross between rummy and pairs. Substance cards are dealt to each player and property cards are placed face up in the centre of the table. The players, in turn, have to get rid of their cards by matching them up correctly with the property cards. Error cards can be used to challenge incorrect pairing. The game ends when one player has played all the cards in his hand. A points system is used, and the winner is the player with the most points at the end of the game.

PROTEINS AS HUMAN FOOD

Authors: F Percival & A H Johnstone

Publisher: Education Division, The Chemical Society, Burlington House, Piccadilly, London W1V 0BN

First published: 1977

Price: £3.00

Type of exercise: manual case study/communication exercise.
Courses for which exercise is suitable: chemistry, biology and health education courses at upper secondary and lower tertiary levels.
Educational aims: to make the participants aware of the nutritional role of proteins and the world protein shortage; to help develop inter-personal, communication, interpretative and decision-making skills.
Number of participants needed: groups of 6 people.
Time required: 1½ hours (minimum).
Contents of package: teacher's guide, resource materials for 6 members of group (A4 booklets).
Special facilities required: none.
Outline of exercise: Proteins as Human Food is a structured communication exercise based on the nutritional role of proteins and the world protein shortage. The participants work in groups of 6, each group comprising a group leader plus 5 members, each of whom is given part of the information needed to hold a meaningful discussion of the protein problem. The object of the exercise is to pool this information, discuss the problem and formulate a possible solution.

PUBLIC INQUIRY PROJECT
Authors: H I Ellington, A G Garrow & J R Muckersie
Publisher: The Association for Science Education, College Lane, Hatfield, Herts AL10 9AA
First published: 1980
Type of exercise: manual simulation game.
Courses for which exercise is suitable: The exercise was specially developed for use in the ASE's Science in Society AO level course, and is suitable for use in science in society and general studies courses at upper secondary and tertiary levels.
Educational aims: to make the participants aware of the complexity and importance to society of a large petrochemical plant and of the social role and procedure of a public inquiry; to cultivate communication and debating skills.
Number of participants needed: The exercise is designed for use with a class of up to 20 pupils or students but can be used with smaller numbers (min 10).
Time required: 1 school double period of roughly 1¼ - 1½ hours.
Contents of package: teacher's guide, student booklets, role cards, maps.
Special facilities required: overhead projector, blank OHP transparencies and felt pens.
Outline of exercise: The Public Inquiry Project is a simplified, shortened version of Point Fields (see page 166). Like the latter, it is based on the hypothesis that a subsidiary of an (imaginary) major oil company plans to build a large petrochemical plant at Point Fields, a coastal site somewhere in the east of Scotland, and that a public inquiry into the proposed scheme has been called. The exercise simulates this public inquiry. The participants are allocated roles supporting or objecting to the planned development,

the various arguments being assessed by an independent 'reporter' who then decides whether or not the scheme should go ahead.

RANDOM
Inquiries: R Lewis, Educational Computer Section, Chelsea College, Pulton Place, London SW6 5RP
First published: 1977
Type of exercise: computer simulation.
Computer language: BASIC or FORTRAN.
Courses for which exercise is suitable: chemistry courses at tertiary level.
Educational aims: to enable students to investigate large numbers of energy exchanges to achieve a good approximation to the Boltzmann Distribution.
Number of participants needed: The exercise can be used either individually or with small groups.
Time required: 1 hour.
Contents of package: teacher's guide, students' notes, computer program.
Special facilities required: graphics terminal or alphanumeric terminal.
Outline of exercise: This package extends a laboratory exercise in which students simulate energy transfer between molecules using dice and counters. In the program the student can specify the size of the array, the number of energy quanta, their distribution, and the type of collision process. It can be used to simulate the change in molecular energy distribution following the mixing of gases with different temperatures.

RATES OF REACTION
Authors: K Shaw & R Edens
Publisher: Schools Council, distributed by Edward Arnold (Publishers) Ltd, London
First published: 1978
Price: part of pack Computers in the Chemistry Curriculum, edited by K Shaw and D Want, price £15.00
Type of exercise: computer simulation.
Computer language: BASIC.
Courses for which exercise is suitable: chemistry courses from middle secondary to lower tertiary level.
Educational aims: to enable students to investigate in detail the factors influencing the rates at which chemical reactions occur.
Number of participants needed: The exercise can be used either individually or with small groups.
Contents of package: teacher's notes, students' leaflets, computer program.
Special facilities required: interactive computer terminal or microcomputer.
Outline of exercise: Students study the effect of reaction conditions, including the particle size of the catalyst, on the rate of decomposition of hydrogen peroxide.

REACTION: CHEMISTRY'S ALPHABET
Authors: R D Stamp & W Harrison
Publisher: Longman Group Ltd, Resources Unit, 9/11 The Shambles, York
First published: 1975
Price: £0.55
Type of exercise: board game.
Courses for which exercise is suitable: chemistry courses at middle to upper secondary level.
Educational aims: to give the participants a basic knowledge of the properties, uses, and reactions of 8 common elements and some of their compounds.
Number of participants needed: 2 - 8 players.
Time required: 1 school period of roughly 35 - 40 minutes.
Contents of package: A4 booklet whose outer sheets (of thin card) are used to prepare the board and other playing materials. The inner sheets (of paper) constitute an 8-page background reader/instructional booklet for the pupils.
Special facilities required: scissors to prepare playing materials, matchstick to mount spinner.
Outline of exercise: The board is in 3 sections: element circuit, compound circuit and reagent pool. Players take turns to take an element token around the element circuit, obeying the instructions on the squares on which they land. On completion of this circuit, they enter the reagent pool and their element token 'reacts' and forms a 'compound' which is then taken round the compound circuit in similar fashion. Part-way round this circuit, the player's token again enters the reagent pool, and reacts to form a new compound. This new compound token then completes the compound circuit. Information leaflets are available to players throughout the game to help them decide whether or not an instruction on a given square is relevant to their element or compound. The winner is the player who completes all 3 circuits first.

RKINET (UNIT ON CHEMICAL REACTION KINETICS)
Author: A W B Aylmer-Kelly
Publisher: Edward Arnold (Publishers) Ltd, London (Chelsea Science Simulation Project)
First published: 1975
Price: £7.25
Type of exercise: computer simulation.
Computer language: BASIC.
Courses for which exercise is suitable: chemistry courses at upper secondary and lower tertiary levels.
Educational aims: to extend students' laboratory experience by enabling them to carry out a wider range of investigations without taking an excessive amount of time; to assist students' understanding of the relationship between a mathematical model and reality.

Number of participants needed: The exercise can be used either individually or with small groups.

Time required: 1½ - 2 hours.

Contents of package: teacher's guide, students' notes, computer program.

Special facilities required: computer terminal or minicomputer.

Outline of exercise: This exercise simulates a number of experiments which are difficult to perform or require special conditions and looks at them from the point of view of reaction kinetics. In each reaction students investigate the effects of changes in temperature and concentration. The simulation also provides experience in selecting investigations which produce meaningful results.

SOLUBILITY AND SUCCESSFUL SEPARATIONS (CHEMISTRY IN ACTION – 9)

Author: N Reid.

Publisher: Heinemann Educational Books, London

First published: 1980

Price: part of a textbook *Chemistry About Us* by A H Johnstone, T I Morrison and N Reid.

Type of exercise: manual case study.

Courses for which exercise is suitable: chemistry courses at middle secondary level (specifically designed for Scottish O Grade).

Educational aims: to make the participants aware that 'solubility' and 'insolubility' are merely relative terms and help them appreciate the application of solubility ideas in solving real-life problems; to help cultivate inter-personal, communication and data-handling skills.

Number of participants needed: groups of 3 or 4.

Time required: roughly 40 minutes.

Contents of package: part of textbook (1 textbook required per participant).

Special facilities required: none.

Outline of exercise: Pupils are given practice in using solubility data. They are then set a series of real-life problems, which they attempt to solve on the basis of the information supplied. These problems range from industrial pollution to a case of poisoning.

TAKE YOUR CHOICE

Author: N Reid

Publisher: Scottish Council for Educational Technology, Dowanhill, Victoria Crescent Road, Glasgow

First published: 1977

Price: £2.00

Type of exercise: manual case study.

Courses for which exercise is suitable: chemistry courses at middle secondary level (designed specifically for Scottish O Grade).

Educational aims: to develop (a) an awareness of relationships between property, use, and personal judgement in relation to fibres; (b) an awareness that decision making is frequently made on the basis of compromise; (c) experience in data handling and communication within a peer group.

Number of participants needed: up to 20 in small groups.

Time required: 1 school double period of roughly 1¼ - 1½ hours.

Contents of package: paperwork.

Special facilities required: none.

Outline of exercise: Participants are given data on 8 unnamed fibres and, working as a group, are asked to choose the best fibres for various tasks. In the second part, an attempt is made to relate property to structure, using water absorption as an illustration.

THE ALKALI INDUSTRY

Authors: K C Campbell, M J Easton & A H Johnstone

Publisher: The University of Glasgow (information available from the authors at the Dept of Chemistry, University of Glasgow)

First published: 1976

Type of exercise: computer simulation.

Computer language: Extended BASIC or FORTRAN IV.

Courses for which exercise is suitable: chemistry courses at upper secondary and lower tertiary levels.

Educational aims: to enable students (a) to make industrial decisions on incomplete information; (b) to work to a budget; (c) to apply chemical knowledge in other disciplines; (d) to make intelligent compromises.

Number of participants needed: groups of 4 - 6 students.

Time required: around 3 hours.

Contents of package: teacher's guide, students' notes, computer program.

Special facilities required: interactive computer terminal.

Outline of exercise: The exercise is based upon the development of a hypothetical find of extensive salt deposits in southern Scotland. Students make decisions on what products to manufacture, where the plant should be sited, what size the plant should be, and how long the plant will take to pay for itself. Students also investigate the environmental implications of their decisions.

THE AMSYN PROBLEM

Authors: F Percival & N Reid

Publisher: Scottish Council for Educational Technology, Dowanhill, Victoria Crescent Road, Glasgow

First published: 1976

Price: £8.50

Type of exercise: simulated case study involving role play.

Courses for which exercise is suitable: chemistry courses at upper secondary and lower tertiary levels.

Educational aims: to develop (a) communications skills in group discussion, logical argument and presentation of a viewpoint; (b) an appreciation of the chemical and social implications relating to an apparently simple process; (c) an appreciation of the relationships between social structures and chemical technology.

Number of participants needed: 9 - 21.
Time required: 2 - 2½ hours.
Contents of package: tape, slides, booklets.
Special facilities required: cassette player, slide projector.
Outline of exercise: Participants are introduced to the problem of the
Amsyn Company through a short tape/slide presentation. They then adopt
roles in a simulation, as they try to solve the problem of the company,
which is faced with closure because of the introduction of new, more
stringent pollution conditions. In the end, they find that there is no easy
solution to the problem.

THE ENERGY PROBLEM
Author: N Reid
Publisher: Scottish Council for Educational Technology, Dowanhill,
Victoria Crescent Road, Glasgow
First published: 1977 (revised 1978, 1980)
Price: £8.00
Type of exercise: simulation game.
Courses for which exercise is suitable: chemistry, general science, science
in society courses at middle and upper secondary level.
Educational aims: to provide (a) knowledge of the energy prospects for
Britain; (b) awareness of the scale of investment required to develop
energy sources, and to encourage a willingness to make choices, and
relate the results to future choices.
Number of participants needed: up to 20 players in 4 groups.
Time required: 1 school double period of roughly 1¼ - 1½ hours.
Contents of package: tape, slides, paperwork.
Special facilities required: cassette player, slide projector. A computerized
version is also available. A *PET* cassette is available from the Computing
Department, Jordanhill College of Education, Southbrae Drive, Glasgow.
Outline of exercise: Pupils compete in groups to plan for the development
of primary energy for Britain for the next half-century. Scores are
awarded according to success.

THE FERTILIZER PROBLEM
Author: N Reid
Publisher: Scottish Council for Educational Technology, Dowanhill,
Victoria Crescent Road, Glasgow
First published: 1977
Price: £1.50
Type of exercise: manual case study.
Courses for which exercise is suitable: chemistry courses at middle
secondary level (designed specifically for Scottish O Grade).
Educational aims: to develop (a) knowledge of the nature of fertilizers in
chemical terms; (b) an awareness of some of the factors influencing choice
of fertilizers; (c) an awareness of the contribution of chemistry to food
production; (d) experience of data handling and communication within
a peer group.

Number of participants needed: up to 20 in groups of 3 or 4.
Time required: 1 school double period of roughly 1¼ - 1½ hours.
Contents of package: paperwork.
Special facilities: none.
Outline of exercise: Pupils are given data on a range of compounds that might be useful in making a compound fertilizer. They work in groups to assess the evidence, and choose the most likely compounds.

THE MANUFACTURE OF SULPHURIC ACID
Authors: R Edens & D Want
Publisher: Schools Council, distributed by Edward Arnold (Publishers) Ltd, London
First published: 1978
Price: part of pack Computers in the Chemistry Curriculum edited by K Shaw and D Want, price £15.00
Type of exercise: computer simulation.
Computer language: BASIC.
Courses for which exercise is suitable: chemistry courses from middle secondary to lower tertiary level.
Educational aims: to present information about a process in such a way that the principles underlying the choice of conditions can be worked out with a consequent improvement in retention and understanding; to show what is meant by 'chemical principles' and how compromise with these is almost inevitable.
Number of participants needed: groups of 3 or 4.
Contents of package: teacher's notes, students' leaflets, computer program.
Special facilities required: interactive computer terminal or microcomputer.
Outline of exercise: Students investigate the conflicting demands of best equilibrium yield, high daily production and profitable manufacture in a sulphuric acid plant.

THE PROTEIN PROBLEM
Author: N Reid (based on an original idea from F Percival)
Publisher: Scottish Council for Educational Technology, Dowanhill, Victoria Crescent Drive, Glasgow
First published: 1977
Price: £2.25
Type of exercise: manual case study.
Courses for which exercise is suitable: chemistry courses at middle and upper secondary level.
Educational aims: to provide (a) knowledge of the basic facts of protein chemistry and the importance of protein in living systems; (b) awareness of the nature of the world protein shortage, and what might be done about it; also, to encourage communication skills and a willingness to argue and discuss issues relating to large-scale problems.
Number of participants needed: up to 24 in groups of 6.
Time required: 1 school double period of roughly 1¼ - 1½ hours.
Contents of package: sheets of paper.

Special facilities required: none.
Outline of exercise: After an introduction to the topic of proteins, pupils are divided into groups of 6, and 6 different sheets are issued. 5 of the sheets provide information on 1 aspect of the world problem, the 6th being a chairman's guide, which contains a series of questions. A highly structured discussion follows.

THE PVC STORY (CHEMISTRY IN ACTION – 2)
Author: N Reid
Publisher: Heinemann Educational Books, London
First published: 1980
Price: part of a textbook *Chemistry About Us* by A H Johnstone, T I Morrison and N Reid.
Type of exercise: manual case study.
Courses for which exercise is suitable: chemistry courses at middle secondary level (specifically designed for Scottish O Grade).
Educational aims: to make the participants aware of the modern rapid growth of plastic production in Britain and appreciate that this growth will probably not continue indefinitely; to cultivate inter-personal, communication and decision-making skills.
Number of participants needed: groups of 3 or 4.
Time required: roughly 40 minutes.
Contents of package: part of textbook (one textbook required for each participant).
Special facilities required: none.
Outline of exercise: Pupils are introduced to the growing use of PVC, and to 3 possible routes to the monomer. They have to select the best route and look at various other aspects of production.

THE SULPHURIC ACID STORY
Author: N Reid
Publisher: Scottish Council for Educational Technology, Dowanhill, Victoria Crescent Road, Glasgow
First published: 1977
Price: £6.00
Type of exercise: manual case study.
Courses for which exercise is suitable: chemistry courses at middle secondary level (designed specifically for Scottish O Grade).
Educational aims: to provide (a) a basic understanding of the salient aspects of sulphuric acid production; (b) an awareness of the changing pattern of industrial production as external conditions change; to encourage a willingness to make decisions on the basis of supplied data; to develop communication skills within a peer group.
Number of participants needed: 9 - 21, in groups.
Time required: 1 school double period of roughly 1¼ - 1½ hours.
Contents of package: tape, booklets and work sheets.
Special facilities required: cassette player.
Outline of exercise: Pupils work in 3 groups to take simulated production

decisions at 3 points in the past, covering a range of 100 years of acid production in Britain. Up-to-date data is provided at the end, and the current data assumes a greater significance in the context of the history of the industry.

THE YOUNG CHEMIST

Authors: U Zoller & J Timor
Publisher: further information from the authors at Division of Teaching in Science and Technology, Technion-Israel Institute of Technology, Technion City, Haifa, Israel
First published: 1980
Type of exercise: board game.
Courses for which exercise is suitable: chemistry courses at middle secondary level.
Educational aims: to provide the participants with the opportunity to exercise the scientific methodology of logical process; to provide a vehicle for the participants to achieve a wide variety of cognitive (eg periodic table, chemical bonding) and non-cognitive (eg co-operation, inquiry) outcomes.
Number of participants needed: no restrictions but minimum of 4.
Time required: 1 school double period of roughly 1¼ - 1½ hours.
Contents of package: a collection of cards (game story, question, clue, information, element, graph); 2 game boards.
Special facilities required: none.
Outline of exercise: The class is divided into groups of 4 within which there are competing pairs who have to pick up question cards in numbered order and discuss (within each pair) the presented question in an attempt to reach the right solution. The answer agreed upon is then compared to that of the opposing pair and the two answers checked against the answer sheet. Clues can be taken in trying to find the correct answer and a scoring system is used on each item. The winning pair (within each group) is the one which scores the most points over the entire game, which ends with the successful independent construction of the first part (18 elements) of the periodic table.

WHAT HAPPENS WHEN THE GAS RUNS OUT?

Authors: F Percival & A H Johnstone
Publisher: Education Division, The Chemical Society, Burlington House, Piccadilly, London W1V 0BN
First published: 1977
Price: £2.50
Type of exercise: manual case study.
Courses for which exercise is suitable: chemistry courses at upper secondary and lower tertiary levels.
Educational aims: to make the participants aware of the problems that will have to be faced by Britain's gas industry over the next 20 - 30 years; to help develop communication, inter-personal, interpretative, problem-solving and decision-making skills.

Number of participants needed: groups of 4 - 8 people.
Time required: 2½ hours (minimum).
Contents of package: teacher's guide; resource materials for participants (A4 sheets).
Special facilities required: none.
Outline of exercise: What Happens When The Gas Runs Out? is a structured case study on the future of Britain's gas industry. Working in co-operative groups, the participants have to study the various options available and formulate an overall policy for the eventual replacement of natural gas.

WHAT IS AN EXPLOSIVE?
Author: N Reid
Publisher: Scottish Council for Educational Technology, Dowanhill, Victoria Crescent Road, Glasgow
First published: 1977
Price: £4.00
Type of exercise: manual case study.
Courses for which exercise is suitable: chemistry courses at middle secondary level (designed specifically for Scottish O Grade).
Educational aims: to develop (a) an understanding of the essential characteristics of explosives; (b) an awareness of the history of explosives; (c) an appreciation of the moral dangers in using explosives and the attitudes represented by Nobel.
Number of participants needed: any number up to 20.
Time required: 1 school double period of roughly 1¼ - 1½ hours.
Contents of package: tape and booklets.
Special facilities required: cassette player.
Outline of exercise: The package uses the development of explosives to discuss the nature of burning and rates of reaction. The modern-day explosive is set in an historical context, and the contribution of Nobel considered. Participants conclude by writing short essays.

Section 3: Biology-based exercises

BIOLOGY CROSSWORDS
Authors: B Eveling & S Hind
Publisher: Sigma Technical Press, 23 Dippons Mill Close, Tettenhall, Wolverhampton WV6 8HH
First published: 1978
Price: £1.25
Type of exercise: set of crossword puzzles.
Courses for which exercise is suitable: biology courses at middle secondary level (specifically designed for English O level course).
Educational aims: to serve as a reinforcement and revision tool covering all aspects of biology included in the above course.
Number of participants needed: can be used with groups of any size or for private study.

Time required: Each crossword requires roughly 30 minutes to complete.
Contents of package: teacher's guide, set of 20 crosswords (1 copy of each). with solutions.
Special facilities required: none.
Outline of exercise: The participants complete the crosswords in the conventional way. The 20 puzzles in the pack cover different branches of general biology, human biology and botany.

BREAKDOWN – A DIGESTION GAME
Authors: R D Stamp & W Harrison
Publisher: Longman Group Ltd, Resources Unit, 9/11 The Shambles, York
First Published: 1975
Price: £0.55
Type of exercise: board game.
Courses for which exercise is suitable: biology courses at middle to upper secondary level.
Educational aims: to help participants understand the process of digestion.
Number of participants needed: 2 - 4 players.
Time required: 1 school double period of roughly 1¼ - 1½ hours.
Contents of package: A4 booklet whose outer sheets (of thin card) are used to prepare the board and other playing materials. The inner sheets (of paper) constitute a 9-page background reader/instructional booklet for the pupils.
Special facilities required: scissors to prepare playing materials, matchsticks to mount spinners.
Outline of exercise: The board represents the food channels in the body. The participants take a food load around the board using a spinner and act on the instructions on the squares as appropriate. The game is finished when all players have left the system.

CIRC
Inquiries: R Lewis, Educational Computing Section, Chelsea College, Pulton Place, London SW6 5PR
First published: 1977
Type of exercise: computer simulation.
Computer language: BASIC or FORTRAN.
Courses for which exercise is suitable: biomedical courses at tertiary level.
Educational aims: to allow the student to discover the properties of the heart as an autoregulator, neural and hormonal controls having been removed.
Number of participants needed: The exercise can be used either individually or with small groups.
Time required: 1 hour.
Contents of package: teacher's guide, students' notes, computer program.
Special facilities required: graphics terminal.
Outline of exercise: This program provides an introduction to the circulatory system using a simple model. Initially the Starling heart-lung

preparation is simulated, and the user can alter variables such as the temperature, fluid resistance, or hydrostatic pressure of blood returning to the heart. The program displays plots of 5 dependent variables (heart rate, arterial pressure, stroke volume, right atrial pressure, flow rate) as functions of 1 or 2 independent variables. The effects of drug addiction (eg adrenalin or barbiturates) and ageing of the preparation can be investigated.

CLASSIFICATION

Authors: R D Stamp & W Harrison
Publisher: Longman Group Ltd, Resources Unit, 9/11 The Shambles, York
First published: 1975
Price: £0.55
Type of exercise: board game.
Courses for which exercise is suitable: biology and general science courses at middle to upper secondary level.
Educational aims: to help the participants understand the classification process for living and non-living things found in nature.
Number of participants required: up to 8 players.
Time required: 1 school double period of roughly 1¼ - 1½ hours.
Contents of package: A4 booklet whose outer sheets (of thin card) are used to prepare the board and other playing materials. The inner sheets (of paper) constitute an 8-page background reader/instruction booklet for the pupils.
Special facilities required: scissors to prepare playing materials.
Outline of exercise: The board is in 4 sections, each representing a part of the living or non-living world. The exercise is in 2 parts. In the first the participants have to arrange a set of cards on *one section* of the board, thus classifying one section of the living or non-living world. In the second, which is competitive, the participants play on all sections of the board, thus gaining an appreciation of how the world can be ordered by using classification procedures.

COEXIST (UNIT ON POPULATION DYNAMICS)

Author: P J Murphy
Publisher: Edward Arnold (Publishers) Ltd, London (Chelsea Science Simulation Project)
First published: 1975
Price: £7.25
Type of exercise: computer simulation.
Computer language: BASIC.
Courses for which exercise is suitable: biology courses at upper secondary and lower tertiary levels.
Educational aims: to demonstrate that an apparently straightforward biological situation, such as the interaction between two species and their environment, is in fact quite complex, although this complexity can be overcome by considering the problem in its component parts; to help

students to design experiments; to illustrate how mathematical models can be used in biology.

Number of participants needed: The exercise can be used either individually, or with small groups.

Time required: 1 - 2 hours.

Contents of package: teacher's guide, students' notes, computer program.

Special facilities required: computer terminal or microcomputer.

Outline of exercise: In this exercise, experiments can be simulated which investigate the effect of changing certain parameters (initial population size, saturation population size, number of offspring, generation time) which have a bearing on the growth rates of non-competing species. It can also be used to carry out a second series of less straightforward 'experiments' on competing species in which the influence of these four parameters and a fifth, the competitive inhibitory factor, can be studied.

COMPETE (UNIT ON PLANT COMPETITION)

Author: M E Leveridge

Publisher: Edward Arnold (Publishers) Ltd, London (Chelsea Science Simulation Project)

First published: 1977

Price: £7.25

Type of exercise: computer simulation.

Computer language: BASIC.

Courses for which exercise is suitable: biology courses at middle and upper secondary level.

Educational aims: to enable students to study interactions between flowering plants; to plan and investigate plant growth experiments.

Number of participants needed: The exercise can be used either individually or with small groups.

Time required: 2 - 3 hours.

Contents of package: teacher's guide, students' notes, computer program.

Special facilities required: interactive computer terminal.

Outline of exercise: This unit combines actual experimental work and use of second-hand data with two computer simulations of plant growth. The first of these simulates monoculture experiments in which the growth of different kinds of plants can be simulated by different students. The second simulation gives students an opportunity to plan an investigation on the interactions between different mixtures of plants.

COUNTERCURRENT SYSTEMS

Author: M E Leveridge

Publisher: Schools Council, distributed by Edward Arnold (Publishers) Ltd, London

First published: 1978

Price: part of a pack Computers in the Biology Curriculum edited by M E Leveridge, price £15.75

Type of exercise: computer simulation.

Computer language: BASIC.

Courses for which exercise is suitable: biology courses at upper secondary and lower tertiary levels.

Educational aims: to demonstrate and explore models of exchangers and multipliers.

Number of participants needed: The exercise can be used either individually or with small groups.

Contents of package: teacher's notes, students' leaflets, computer program.

Special facilities required: interactive computer terminal or microcomputer.

Outline of exercise: The computer sets up simple models of a heat exchanger and a countercurrent multiplier which can then be investigated by students.

DYE

Inquiries: R Lewis, Educational Computing Section, Chelsea College, Pulton Place, London SW6 5PR

First published: 1977

Type of exercise: computer simulation.

Computer language: BASIC or FORTRAN.

Courses for which exercise is suitable: biomedical courses at tertiary level.

Educational aims: to make students aware of the principles involved in the indicator method for measuring cardiac output.

Number of participants needed: The exercise can be used either individually or with small groups.

Time required: 1 hour.

Contents of package: teacher's guide, students' notes, computer program.

Special facilities required: graphics terminal.

Outline of exercise: There are two programs which simulate the injection of an indicator dye into the blood stream of a patient. The first is relatively simple and the student learns something of exponential plots and time constants. The second is more realistic and the student performs semi-log plots to determine the patient's cardiac output.

ENZKIN (UNIT ON ENZYME KINETICS)

Author: M T Heydeman

Publisher: Edward Arnold (Publishers) Ltd, London (Chelsea Science Simulation Project)

First published: 1977

Price: £7.25

Type of exercise: computer simulation.

Computer language: BASIC.

Courses for which exercise is suitable: biology and biochemistry courses at upper secondary and lower tertiary levels.

Educational aims: to enable students to grasp the fundamentals of enzyme kinetics; to encourage ingenuity in dealing with an unknown system; to provide experience of data handling and interpretation; to introduce students to some of the more complex situations in enzyme kinetics such as substrate inhibition, co-operativity, and the influence of a cofactor.

Number of participants needed: The exercise can be used either
individually or with small groups.
Time required: 2 - 3 hours.
Contents of package: teacher's guide, students' notes, computer program.
Special facilities required: interactive computer terminal or microcomputer.
Outline of exercise: The exercise gives practice in planning an experiment,
rather than simply interpreting the results, through a series of simulated
experiments on enzyme kinetics.

EVOLUT (UNIT ON NATURAL SELECTION)
Author: S McCormick
Publisher: Edward Arnold (Publishers) Ltd, London (Chelsea Science
Simulation Project)
First published: 1975
Price: £7.25
Type of exercise: computer simulation.
Computer language: BASIC
Courses for which exercise is suitable: biology courses at middle and
upper secondary level.
Educational aims: to enable the student (a) to understand the production
of adaptions by the action of selection on random variables; (b) to
understand adaption to specific environmental conditions in relation to
survival value; (c) to gain experience in manipulating models of selection
acting on populations.
Number of participants needed: The exercise can be used either
individually or with small groups.
Time required: 1½ - 2 hours.
Contents of package: teacher's guide, students' notes, computer program.
Special facilities required: interactive computer terminal or microcomputer.
Outline of exercise: The first part of the simulation uses a model whose
use depends on an understanding of the principles of the theory of
evolution, the mechanism of natural selection and variation. In the second
part, the student is introduced to a model with which he can experiment
with selective forces acting in different ways. He can thus change and
direct fluctuations in gene frequency within a gene pool of specified
population. The student is introduced to the computer model by using
an experimental pea model.

HUMAN ENERGY EXPENDITURE
Author: J Denham
Publisher: Schools Council, distributed by Edward Arnold (Publishers)
Ltd, London
First published: 1978
Price: part of a pack Computers in the Biology Curriculum edited by
M E Leveridge, price £15.75
Type of exercise: computer simulation.
Computer language: BASIC.

Courses for which exercise is suitable: biology courses at lower and middle secondary level.

Educational aims: to allow students to explore energy requirements in relation to activity, sex and mass without doing any calculation.

Number of participants needed: The exercise can be used either individually or with small groups.

Contents of package: teacher's notes, students' leaflets, computer program.

Special facilities required: interactive computer terminal or microcomputer.

Outline of exercise: The computer program calculates human energy requirements for different activities which can be specified by students. It contains data for men and women undertaking 72 different activities.

INHERITANCE

Author: M E Leveridge

Publisher: Schools Council, distributed by Edward Arnold (Publishers) Ltd, London

First published: 1978

Price: part of a pack Computers in the Biology Curriculum edited by M E Leveridge, price £15.75

Type of exercise: computer simulation.

Computer language: BASIC.

Courses for which exercise is suitable: biology courses from middle secondary to lower tertiary level.

Educational aims: to supplement students' breeding investigations carried out in the normal laboratory; to enable students to plan crosses and analyse results carefully.

Number of participants needed: The exercise can be used either individually or with small groups.

Contents of package: teacher's notes, students' leaflets, computer program.

Special facilities required: interactive computer terminal or microcomputer.

Outline of exercise: In this computer simulation, students use 4 simulated breeding investigations for the study of genetics. 3 of the species studied are animals (Drosophila, mice and man) and 1 is a plant (tomato). A 5th program contains a simple model of multifactorial inheritance in a hypothetical situation.

INVASION (MICROBES)

Authors: R D Stamp & W Harrison

Publisher: Longman Group Ltd, Resources Unit, 9/11 The Shambles, York

First published: 1975

Price: £0.55

Type of exercise: board game.

Courses for which exercise is suitable: biology and general science courses at middle to upper secondary level.

Educational aims: The game simulates the wars that are constantly being fought to maintain the health of our bodies. It is designed to help pupils learn how it is possible for the body to ward off the attacks by various disease-causing microbes by deploying a range of different defence mechanisms.

Number of participants needed: 2 - 8 players.

Time required: 1 school period of roughly 35 - 40 minutes for each section of the game to be covered by a given pupil. (The 4 sections can be played simultaneously by up to 8 pupils).

Contents of package: A4 booklet whose outer sheets (of thin card) are used to prepare the board and other playing materials. The inner sheets (of paper) constitute a 12-page background reader/instructional booklet for the pupils.

Special facilities required: scissors to prepare board and playing pieces, matchstick for mounting the spinner.

Outline of exercise: Pupils play in pairs, each pair using one sector of the board which represents a part of the body (lung, gut, sex organ and skin). Each pupil chooses an opponent, a sector of the board and a role (defender or attacker) before play commences. The game is in 2 parts: in the first, the attacker and defender try to colonize their section of the board and draw up battle lines; in the second, the attacker tries to occupy as many areas of the battlefield as possible, while the defender tries to eliminate the attacks. A scoring system is used to decide whether the attack has been successful or whether the potential disease has been overcome.

LINKOVER (UNIT ON GENETIC MAPPING)

Author: P J Murphy

Publisher: Edward Arnold (Publishers) Ltd, London (Chelsea Science Simulation Project)

First published: 1975

Price: £7.25

Type of exercise: computer simulation.

Computer language: BASIC.

Courses for which exercise is suitable: biology courses at upper secondary and lower tertiary levels.

Educational aims: to improve students' comprehension of linkage and genetic mapping; to allow students to plan and execute a programme of associated experiments.

Number of participants needed: The exercise can be used either individually or with small groups.

Time required: 3 hours of computer sessions plus 2½ hours of associated classwork.

Contents of package: teacher's guide, students' notes, computer program.

Special facilities required: interactive computer terminal or microcomputer.

Outline of exercise: The first stage of the computer investigation involves the selection of gene combinations either alphabetically or otherwise until the sequence of the genes in the linkage group has been established. The second stage involves the selection of adjacent genes so that the distances separating the genes can be accurately calculated. The students are then involved in producing an accurate genetic map of the single chromosome of the 'electronic species' under investigation.

NUTRITION
Authors: R D Stamp & W Harrison
Publisher: Longman Group Ltd, Resources Unit, 9/11 The Shambles, York
First published: 1975
Price: £0.55
Type of exercise: board game.
Courses for which exercise is suitable: biology courses at middle to upper secondary level.
Educational aims: to help participants understand the energy chain, which converts the sun's energy into food; to make the participants aware of the types and quantities of foodstuffs required to provide man with a balanced diet; to highlight the world's present food crisis.
Number of participants needed: 2 - 4 players.
Time required: 2 school periods of roughly 35 - 40 minutes each (1 period for each game).
Contents of package: A4 booklet whose outer sheets (of thin card) are used to prepare the boards and other playing materials. The inner sheets (of paper) constitute a 6-page background reader/instructional booklet for the pupils.
Special facilities required: scissors to prepare the playing materials.; matchstick for mounting the spinner.
Outline of the exercise: The package consists of 2 games. The first, The Energy Chain, involves the participants in progressing to the top of a pyramid using a spinner to overcome obstacles on the way. They each start with the same amount of energy, but end up with different amounts, depending on how successfully they climbed the pyramid. In the second, Facing Famine, the participants have to produce a balanced diet in both the developed world and the undeveloped world. They are asked to fill in the nutritional values of all the food in the meals described on various parts of the board on their daily diet sheets. The game ends when all players have had the opportunity to collect nutrients from each meal on the board.

OPERON
Inquiries: R Lewis, Educational Computing Section, Chelsea College, Pulton Place, London SW6 5PR
First published: 1977
Type of exercise: computer simulation.
Computer language: BASIC or FORTRAN.
Courses for which exercise is suitable: biology courses at tertiary level.
Educational aims: to allow students to combine genetic elements at will.
Number of participants needed: The exercise can be used either individually or with small groups.
Time required: 1 hour.
Contents of package: teacher's guide, students' notes, computer program.

Special facilities required: graphics terminal or alphanumeric terminal.
Outline of exercise: This package provides an introduction to the
Jacob-Monod model of genetic induction. Various options allow the
student to select one, or the cross of two genetic strains, or to create his
own by specifying the order of genes within the operon and the content
of the genes. The student can then make a simulated assay for enzymic
activity, with or without inducer, for various times of sampling, total
assay and introduction of inducer.

POLY

Inquiries: R Lewis, Educational Computing Section, Chelsea College,
Pulton Place, London SW6 5PR
First published: 1977
Type of exercise: computer simulation.
Computer language: BASIC or FORTRAN.
Courses for which exercise is suitable: biology courses at tertiary level.
Educational aims: to reinforce teaching on polygenic inheritance.
Number of participants needed: The exercise can be used either
individually or with small groups.
Time required: 1 hour.
Contents of package: teacher's guide, students' notes, computer program.
Special facilities required: graphics terminal.
Outline of exercise: The package models a simple situation of polygenic
inheritance. The student controls the number of relevant loci (up to 5),
the type of dominance (classical or co-dominance), the phenotypic effect
of each allele and the parental genotypes. Any number of offspring are
then produced from the specified mating, and a histogram is built up
showing the number of organisms in each phenotype class. The model
can also be used to set simple statistical exercises using random
distributions.

POND ECOLOGY

Authors: J A Tranter & M E Leveridge
Publisher: Schools Council, distributed by Edward Arnold (Publishers)
Ltd, London
First published: 1978
Price: part of a pack Computers in the Biology Curriculum edited by
M E Leveridge, price £15.75
Type of exercise: computer simulation.
Computer language: BASIC.
Courses for which exercise is suitable: biology courses from middle
secondary to lower tertiary level.
Educational aims: to enable students to study the interactions between
the trophic levels and some of the effects of man upon them.
Number of participants needed: The exercise can be used either
individually or with small groups.
Contents of package: teacher's notes, students' leaflets, students' notes,
computer program.

Special facilities required: interactive computer terminal or microcomputer.
Outline of exercise: This computer simulation sets up a model ecosystem which allows the effects of fishing and pollution on the numbers of phytoplankton, herbivores and fish to be investigated.

PREDATOR-PREY RELATIONSHIPS
Author: J Denham
Publisher: Schools Council, distributed by Edward Arnold (Publishers) Ltd, London
First published: 1978
Price: part of a pack Computers in the Biology Curriculum edited by M E Leveridge, price £15.75
Type of exercise: computer simulation.
Computer language: BASIC.
Courses for which exercise is suitable: biology courses at upper secondary and lower tertiary levels.
Educational aims: to enable students to study some of the effects of the interactions between predators and prey.
Number of participants needed: The exercise can be used either individually or with small groups.
Contents of package: teacher's notes, students' leaflets, computer program.
Special facilities required: interactive computer terminal or microcomputer.
Outline of exercise: A simple model of the relationship between predator and prey is set up and investigated. The effects of the interactions upon the sizes of predator and prey populations are studied by setting up a model of an ecosystem containing 1 prey species and 1 predator species.

SELECT
Inquiries: R Lewis, Educational Computing Section, Chelsea College, Pulton Place, London SW6 5PR
First published: 1977
Type of exercise: computer simulation.
Computer language: BASIC or FORTRAN.
Courses for which exercise is suitable: biology courses at tertiary level.
Educational aims: to reinforce teaching on population genetics.
Number of participants needed: The exercise can be used either individually or with small groups.
Time required: 1 hour.
Contents of package: teacher's guide, students' notes, computer program.
Special facilities required: graphics terminal or alphanumeric terminal.
Outline of exercise: This package models natural selection in action for a one gene/two allele system. The student can control the following parameters: relative viability of the genotypes, mutation rates, migration rates, size and natural composition of the population and the breeding system in operation in order to investigate how the population changes through any number of generations. The model is stochastic for small

populations and deterministic for large ones.

STATISTICS FOR BIOLOGISTS

Author: M E Leveridge
Publisher: Schools Council, distributed by Edward Arnold (Publishers) Ltd, London
First published: 1978
Price: part of a pack Computers in the Biology Curriculum edited by M E Leveridge, price £15.75
Type of exercise: computer simulation.
Computer language: BASIC.
Courses for which exercise is suitable: biology courses at upper secondary and lower tertiary levels.
Educational aims: to help students appreciate the statistics used in biology.
Number of participants needed: The exercise can be used either individually or with small groups.
Contents of package: teacher's notes, students' leaflets, computer program.
Special facilities required: interactive computer terminal or microcomputer.
Outline of exercise: The computer simulates results of biological experiments, the data from which is used to assist in the development of students' understanding of statistical techniques.

THE DEAD RIVER

Author: E N Swinerton
Publisher: Union Printing Co Inc, 17 W Washington Street, Athens, Ohio 45701, USA
First published: 1973
Price: $14.00 (reductions available for orders of 10 or more)
Type of exercise: manual role-playing simulated case study.
Courses for which exercise is suitable: biology, ecology, environmental studies, social studies and general studies courses at middle secondary to lower tertiary level.
Educational aims: to make the participants aware of the complexities of water resource problems and the methods used to solve them; to help develop inter-personal, communication, decision-making and problem-solving skills.
Number of participants needed: 10 - 30.
Time required: 2 - 3 hours (minimum).
Contents of package: teacher's manual, sets of team guide books, team name plates.
Special facilities required: none.
Outline of exercise: The Dead River simulates a water pollution problem. Each participant is given a role to play in planning to clean up a river and forms part of a team ('taxpayers', 'developers', 'recreation association members', 'scientists' and 'others'). The various teams try to establish

the quality standards they think are important for their particular uses of the river, after which they try to agree on a quality standard acceptable to and within the financial means of all the groups.

THE GREAT BLOOD RACE
Authors: R D Stamp & W Harrison
Publisher: Longman Group Ltd, Resources Unit, 9/11 The Shambles, York
First published: 1975
Price: £0.55
Type of exercise: board game.
Courses for which exercise is suitable: biology and general science courses at middle to upper secondary level.
Educational aims: to help the participants appreciate the composition and circulation of blood in the human body, the main organs it visits and the reasons why, and some of the things that can go wrong with the circulation system.
Number of participants needed: 2 - 6 players.
Time required: 1 school period of roughly 35 - 40 minutes.
Contents of package: A4 booklet whose outer sheets (of thin card) are used to prepare the board and other playing materials. The inner sheets (of paper) constitute an 8-page background reader/instructional booklet for the pupils.
Special facilities required: scissors to prepare board and playing pieces; personal counters (1 per pupil), 2 matchsticks for mounting the spinners.
Outline of exercise: The board is designed to represent the blood circulation system in the human body and the exercise is really a race around the circulation system. The game starts by pupils taking turns with a spinner in order to collect together the 5 components of whole blood. Once these have been acquired, the pupil can enter the board game proper. The rate of progress round the track representing the circulation system is controlled by a 6-sided spinner and is helped or hindered by various chance and other factors. The hindering factors occur regularly and confront the pupils with problems they can only solve if they know certain facts about human blood. The winner of the game is the first pupil to complete a single circuit of the board.

THE RIDPEST FILE
Publisher: BP Educational Service, PO Box 5, Wetherby, West Yorkshire LS23 7EH
First published: 1977
Price: £12.96
Type of exercise: manual simulated case studies.
Courses for which exercise is suitable: science courses at upper secondary and tertiary levels.
Educational aims: to enable the students to participate in the experience of decision making in the fields of biology and environmental studies.

Number of participants needed: up to 24.

Time required: up to 4 hours for each case study.

Contents of package: teacher's guide, audio cassette, filmstrip, information cards, newspapers, work sheets.

Special facilities required: cassette player, filmstrip viewer.

Outline of exercise: The package consists of 4 simulated case studies, dealing with different aspects of pests and pest control. The material in each case study is adaptable for use with different ability groups. The participants use the resource material and the work sheets in order to make important decisions on crop protection and public health. The titles of the individual case studies are as follows:

1. Plants in the wrong place
2. Potato famine
3. Insects at home and abroad
4. Help the harvest.

TRANSPIRATION

Authors: M E Leveridge & J Pluck

Publisher: Schools Council, distributed by Edward Arnold (Publishers) Ltd, London

First published: 1978

Price: part of a pack Computers in the Biology Curriculum edited by M E Leveridge, price £15.75

Type of exercise: computer simulation.

Computer language: BASIC.

Courses for which exercise is suitable: biology courses from middle secondary to lower tertiary level.

Educational aims: to allow students to carry out investigations into the environmental effects of transpiration.

Number of participants needed: The exercise can be used either individually or with small groups.

Contents of package: teacher's notes, students' leaflets, computer program.

Special facilities required: interactive computer terminal or microcomputer.

Outline of exercise: The effects of the environment on water loss from plants are modelled and studied.

TRANSPORT IN PLANTS

Authors: R D Stamp & W Harrison

Publisher: Longman Group Ltd, Resources Unit, 9/11 The Shambles, York

First published: 1975

Price: £0.55

Type of exercise: board game.

Courses for which exercise is suitable: biology and general science courses at middle to upper secondary level.

Educational aims: to help the participants appreciate the food requirements of plants, how this food is made from raw materials by the process of photosynthesis and how it is transported to the roots for nourishment (the potato tuber is used as the example).

Number of participants needed: 4 players.

Time required: 1 school period of roughly 35 - 40 minutes.

Contents of package: A4 booklet whose outer sheets (of thin card) are used to prepare the board and other playing materials. The inner sheets (of paper) constitute an 8-page background reader/instructional booklet for the pupils.

Special facilities required: scissors to prepare board and playing pieces, 2 matchsticks for mounting the spinners.

Outline of exercise: The game is divided into 2 sections. In the first, the pupils gather together water, mineral salts and carbon dioxide (being the raw materials) and move them in turn along the board from the soil to the palisade cells of the leaf so that photosynthesis can take place. The various factors governing the rate of uptake of these materials are represented as hazard or bonus squares on this part of the board. In the second part, the object is to transport the food (sucrose) from the leaf to the tuber as quickly as possible. Again the various factors which affect the rate of uptake of sucrose by the tuber are represented by hazard and bonus squares on the board. The winner is the first player to reach the tuber with the food.

VITAMINS

Author: P F Ploutz

Publisher: Union Printing Co Inc, 17 W Washington Street, Athens, Ohio 45701, USA

First published: 1975

Price: $10.00 (reductions available for orders of 10 or more)

Type of exercise: card game.

Courses for which exercise is suitable: biology, nutritional science, health education and general studies courses at secondary and lower tertiary levels.

Educational aims: to help the participants learn about vitamins, their sources and their metabolic functions.

Number of participants needed: 2 - 6.

Time required: This varies according to the number of players and their age, but is typically of the order of 20 - 30 minutes for the basic game.

Contents of package: 33 'vitamin' cards, 33 'source' cards, 33 'function' cards, 22 'disorder' cards, link key chart, spinner, instruction booklet.

Special facilities required: none.

Outline of exercise: Players attempt to form vitamin links by matching vitamins with their sources and functions, the winner being the first player to form 3 valid vitamin/source/function links. Variations of play are included to provide additional open-ended activity.

Section 4: Other exercises with some science or technology content

DENTAL HEALTH PROJECT

Authors: F Percival & H I Ellington
Publisher: The Association for Science Education, College Lane, Hatfield, Herts AL10 9AA
First published: 1980
Type of exercise: manual simulation game.
Courses for which exercise is suitable: The exercise was specially developed for use in the ASE's Science in Society AO level course and is suitable for use in science in society, general studies, sociology and health education courses at upper secondary and tertiary levels.
Educational aims: to make the participants aware of the arguments for and against fluoridation of public water supplies and of the procedure by which decisions regarding public health measures are reached at local government level; to cultivate debating and decision-making skills.
Number of participants needed: The exercise is designed for use with a class of up to 20 pupils or students, but can be used with smaller numbers (min 10).
Time required: 1 school double period of roughly 1¼ - 1½ hours.
Contents of package: teacher's guide, role sheets.
Special facilities required: none.
Outline of exercise: The Dental Health Project is a simplified version of Fluoridation? (see page 194). It is based on the hypothesis that an area health authority has recommended that the local water supply should be fluoridated, and takes the form of a simulated meeting of the County Council, who have to make the final decision on the issue. The participants all take the role of county councillors, and are given individual role sheets which state their initial position on the fluoridation issue (for or against) and outline the points that they should make at the meeting, which ends with a free vote.

EKOFISK — ONE OF A KIND

Authors: The main educational material in the package was written by H I Ellington and E Addinall
Publisher: Phillips Petroleum Company Europe-Africa, Portland House, Stag Place, London SW1E 5DA
First published: 1980
Price: supplied to teachers free on request
Type of exercise: multi-disciplinary multi-project and resource pack.
Courses for which exercise is suitable: The package contains resource material, projects and case studies for use in teaching a wide range of subjects (including physics, chemistry, science in society and general studies) from upper primary, through secondary, to lower tertiary level.
Educational aims: to explain what goes on in the North Sea oil and gas industries; to demonstrate the application of physics, chemistry and

other academic subjects to an important real-life situation.

Number of participants needed: The various projects and case studies in the package can be used with any size of class simply by making the appropriate number of photocopies of the relevant class sheet(s).

Time required: Most of the projects and case studies can be completed in a standard school period of roughly 40 minutes.

Contents of package: teacher's guide, set of class sheets (in form of photocopy masters), set of class readers, illustrated booklet, set of posters, glossary of terms. A 30-minute 16mm colour film, *Ekofisk — One of a Kind*, is also available on free loan from Phillips Petroleum.

Special facilities required: access to photocopier (for duplicating class sheets), 16mm film projector (if film is to be shown).

Outline of exercise: Ekofisk — One of a Kind is a form of multi-disciplinary multi-project pack, containing background information on all aspects of the North Sea oil and gas industries plus projects and case studies designed for use in teaching a wide range of academic subjects. The package includes a number of case studies and projects specifically designed for use in teaching physics and chemistry at secondary and lower tertiary levels.

FLUORIDATION?

Authors: F Percival & H I Ellington

Publisher: The Institution of Electrical Engineers, Savoy Place, London WC2R 0BL

First published: 1980

Price: £7.00 (inc p & p)

Type of exercise: manual simulation game.

Courses for which exercise is suitable: science in society, general studies, sociology and health education courses at upper secondary and tertiary levels.

Educational aims: to make the participants aware of the arguments for and against fluoridation of public water supplies and of the procedure by which decisions regarding public health measures are reached at local government level; to cultivate interpretative, communication, debating and decision-making skills.

Number of participants needed: optimum number 18 (min 13; max 24).

Time required: 1¼ - 3 hours, depending on the level at which the exercise is being used.

Contents of package: teacher's guide, introductory booklets, briefing booklets for various roles.

Special facilities required: overhead projector, blank OHP transparencies and felt pens.

Outline of exercise: Fluoridation? is a role-playing simulation exercise. It is based on the hypothesis that an area health authority is considering the principle of fluoridation of the public water supply, and takes the form of a simulated public meeting called by one of the community health councils to discuss the question. The participants are divided into three groups, namely supporters of fluoridation, objectors and 'neutrals' — members of the local health council who listen to the various arguments presented and then decide whether or not to support fluoridation when the issue is discussed at a higher level.

GEOLOGIC TIME CHART GAME
Author: P F Ploutz
Publisher: Union Printing Co Inc, 17 W Washington Street, Athens, Ohio 45701, USA
First published: 1972
Price: $16.50 (reductions available for orders of 10 or more)
Type of exercise: board game.
Courses for which exercise is suitable: geology, biology and general science courses at secondary and lower tertiary levels.
Educational aims: to help the participants to learn the eras and periods of geological time, and make them aware of how species survive or disappear.
Number of participants needed: 2 - 6.
Time required: This varies according to the number of players and their age; typically about 40 minutes.
Contents of package: board, 6 player tokens, 100 population chips, 32 'Evolution' cards, 26 'Chance' cards, die, instructions (on box).
Special facilities required: none.
Outline of exercise: Play proceeds from the Proterozoic Era, when the simplest of life is thought to have originated, through the Palaeozoic, Mesozoic and Cenozoic Eras. Animals are introduced in the appropriate periods of each era and compete with each other and with the environment for survival.

LAB APPARATUS
Author: P F Ploutz
Publisher: Union Printing Co Inc, 17 W Washington Street, Athens, Ohio 45701, USA
First published: 1975
Price: $9.00 (reductions available for orders of 10 or more).
Type of exercise: card game.
Courses for which exercise is suitable: science courses at secondary level.
Educational aims: to help participants identify (by name and appearance) 100 common laboratory apparatus items, become familiar with their use, and become aware of safety precautions associated with their use.
Number of participants needed: normally 2 - 6, although up to 10 can play.
Time required: This varies according to the number of players and their age, but the game is designed to be completed in a single school period of roughly 40 minutes.
Contents of package: 100 cards illustrating lab apparatus, 50 function cards, 50 safety question cards, spinner, instruction/rule booklet.
Special facilities required: none.
Outline of exercise: Players identify lab equipment, then answer questions

about its use and safety. Function and safety cards put players in situations where they can win or lose points. Chance factors add points for good lab technique.

NORTH SEA CHALLENGE
Author: M Lynch, Bath University, School of Education
Publisher: BP Educational Service, PO Box 5, Wetherby, West Yorkshire LS23 7EH
First published: 1975
Price: £12.96
Type of exercise: manual simulated case study.
Courses for which exercise is suitable: science courses at upper secondary and tertiary levels.
Educational aims: to enable the students to participate in the experience of decision making in an area of high technology.
Number of participants needed: 16 - 20.
Time required: up to 3 hours for each case study.
Contents of package: teacher's guide, audio cassette, filmstrip, OHP transparencies, briefing sheets, data cards.
Special facilities required: cassette player, OHP, filmstrip viewer.
Outline of exercise: The package consists of 3 simulated case studies entitled Strike, Slick and Impact. In the first, the participants are faced with an oil strike in the North Sea. Their task is to decide how best to develop the field and get the oil ashore (the decisions being made on economic, environmental, technical and geographical grounds). In Slick, the participants, working in groups, have first to decide on how best to protect an area from a possible oil spill, and then deal with the problems of a 100 tonne oil spill. The final case study, Impact, deals with the implications of industrial development for a small Scottish community; in particular, groups of participants have to prepare cases for and against the proposed development of a steel platform construction yard, and then debate the issue.

OFFSHORE OIL BOARD GAME
Authors: H I Ellington & E Addinall
Publisher: The Association for Science Education, College Lane, Hatfield, Herts AL10 9AA
First published: 1980
Type of exercise: board game.
Courses for which exercise is suitable: The exercise was specially developed for use in the ASE's Science in Society AO level course, and is suitable for use in science in society and general studies courses at middle and upper secondary level.
Educational aims: to make the participants aware of the process by which offshore oilfields are discovered and brought into production and of the high-risk nature of the industry; to help develop the ability to make

decisions under pressure in a rapidly changing situation.

Number of participants needed: The exercise is designed for use with a class of up to 20 pupils working in groups of up to 5.

Time required: 1 school double period of roughly 1¼ - 1½ hours.

Contents of package: teacher's guide, 4 playing sets of the game.

Special facilities required: none.

Outline of exercise: The Offshore Oil Board Game is an educational version of North Sea, a family board game developed by the authors for Shell (UK) Ltd in 1975. The participants take the role of oil companies competing to discover and develop offshore oilfields as quickly as possible, the winner being the first player to bring a field into production and collect sufficient revenue to pay off his/her loans. The game is designed for 2, 3 or 4 players plus a non-playing banker.

QUERIES 'N THEORIES

Authors: L E Allen, Peter Kugel & Joan Ross

Publisher: Science Systems Ltd, 173 Southampton Way, London SE5 7EJ

First published: 1970

Type of exercise: board game.

Courses for which exercise is suitable: science courses at upper secondary and tertiary level.

Educational aims: to help the participants develop an understanding of the basic scientific method — that of formulating theories and testing them.

Number of participants needed: minimum of 2 players.

Time required: as required but a minimum of 1 school double period of roughly 1¼ - 1½ hours is recommended.

Contents of package: game booklet (54pp), board and playing materials.

Special facilities required: none.

Outline of exercise: In the present context, this game is a simulation exercise which deals with the scientific method (it can also be used in the teaching of generative grammars). Queries 'n Theories is a series of games of increasing complexity. Basically they are exercises in formal logic in which one of the participants sets a problem (by defining a language/ theory) and the others have to 'understand' the problem by taking turns to construct 'queries'. Each 'query' is adjudicated before the next is constructed. In this way the participants gradually build up an understanding of the problem set. This is a sophisticated exercise, the time required to play it being dependent on the ability of the participants and the level of play chosen.

SAFETY SNAKES AND LADDERS

Author: A Armstrong

Publisher: Sigma Technical Press, 23 Dippons Mill Close, Tettenhall, Wolverhampton WV6 8HH

First published: 1978

Price: £1.25

Type of exercise: board game.

Courses for which exercise is suitable: science courses at lower to middle secondary level.

Educational aims: to help children just beginning laboratory work to learn effective safety techniques.

Number of participants needed: 2 - 4.

Time required: roughly 30 minutes.

Contents of package: board, die, player tokens.

Special facilities required: none.

Outline of exercise: The game is played in exactly the same way as conventional snakes and ladders, the players moving up ladders and down snakes when they land on squares respectively describing good and bad laboratory practices.

389/SCIENCE CONCEPTS

Authors: P F Ploutz & M Hawkins

Publisher: Union Printing Co Inc, 17 W Washington Street, Athens, Ohio 45701, USA

First published: 1977

Price: $8.00 (reductions available for orders of 10 or more)

Type of exercise: card game.

Courses for which exercise is suitable: science courses at secondary and lower tertiary levels.

Educational aims: to help the participants become familiar with some of the most important concepts in the various life, earth and physical sciences.

Number of participants needed: 2 - 6.

Time required: This varies according to the number of players and their age, but the game is designed to be completed in a single school period of roughly 40 minutes.

Contents of package: 100 Life Science Concept cards, 100 Earth Science Concept cards, 100 Physical Science Concept cards, 34 'Mind Bender' cards, 33 Scientist cards, 33 'Goof' cards, 6 player tokens, instruction booklet.

Special facilities required: none.

Outline of exercise: Science Concepts deals with 16 science topics found in most American elementary/junior high school programmes:

> *Life:* Plants; Animals; Human Body; Micro-Organisms; Ecology.
>
> *Earth:* Astronomy; Meteorology; Oceanography; Earth.
>
> *Physical:* Matter; Energy; Heat-Light-Sound; Magnetism/Electricity; Simple Machines.

Additional features include 33 'Famous Scientists', 33 'Goof' cards (for a 'fun' element) and 34 'Mind-Bender' cards, presenting contemporary science problems (endangered species, energy shortage, space travel, population control, etc). Students collect the various concepts, then 'declare' their interest in a career as an earth, life or physical scientist.

SCIENCE SENSE
Authors: R D Stamp & W Harrison
Publisher: Longman Group Ltd, Resources Unit, 9/11 The Shambles, York
First published: 1975
Price: £0.55
Type of exercise: board game.
Courses for which exercise is suitable: biology, chemistry, general science and physics courses at middle to upper secondary level.
Educational aims: to make the participants aware of the dangers of working in a science laboratory; to make them aware of a commonsense code of safety and procedures and hence help them avoid accidents and mistakes.
Number of participants needed: 2 - 6 players.
Time required: 1 school period of roughly 35 - 40 minutes.
Contents of package: A4 booklet whose outer sheets (of thin card) are used to prepare the board and other playing materials. The inner sheets (of paper) constitute a 4-page background reader/instructional booklet for the pupils.
Special facilities required: scissors to prepare playing materials, matchsticks to mount the spinners.
Outline of exercise: The participants progress around the board using a spinner and act on the instructions on each square on which they land. The winner is the player who completes the circuit first.

SPACE COLONY
Author: J Johnson
Publisher: Teaching Aids Co, 925 South 300 West, Salt Lake City, Utah 84101, USA
First published: 1977
Price: $14.95
Type of exercise: board game.
Courses for which exercise is suitable: general science and general studies courses at lower and middle secondary level.
Educational aims: to introduce the participants to the problems that would be encountered by a space colony trying to establish itself on Mars; to make them aware that survival depends both on good planning and on good luck.
Number of participants needed: 2 - 6.
Time required: 1 hour (minimum).
Contents of package: board, 6 space colony cards, 30 successful experiment cards, 30 accidental failure cards, 12 expedition cards, oxygen, food and water coupons, 90 modules, 12 markers, 2 dice, instruction booklet.
Special facilities required: none.
Outline of exercise: The exercise simulates the problems likely to be faced by an expedition to Mars. Several space vehicles are sent under separate

commanders to establish colonies. Each player represents the formation of a separate colony and his principal objective is survival. Players must ensure that they have enough food, water and oxygen and have to overcome both natural and mechanical disasters.

THE AIR ABOUT US (THE BALLOON RACE)

Authors: R D Stamp & W Harrison
Publisher: Longman Group Ltd, Resources Unit, 9/11 The Shambles, York
First published: 1975
Price: £0.55
Type of exercise: board game.
Courses for which exercise is suitable: general science courses at middle to upper secondary level.
Educational aims: to help the participants understand the complex nature of the earth's atmosphere.
Number of participants needed: 2 - 5 players.
Time required: 1 school double period of roughly 1¼ - 1½ hours.
Contents of package: A4 booklet whose outer sheets (of thin card) are used to prepare the board and other playing materials. The inner sheets (of paper) constitute a 12-page background reader/instructional booklet for the pupils.
Special facilities required: scissors to prepare playing materials, matchstick to mount spinner.
Outline of exercise: The game is based on a hot air balloon race. The progress of each participant is determined principally by knowledge of the atmosphere (the participants have to answer a series of questions relating to the basic properties of the atmosphere). A spinner is used to simulate chance factors and the winner of the race is the winner of the game.

THE WATER CYCLE

Authors: R D Stamp & W Harrison
Publisher: Longman Group Ltd, Resources Unit, 9/11 The Shambles, York
First published: 1975
Price: £0.55
Type of exercise: board game.
Courses for which exercise is suitable: general science courses at middle to upper secondary level.
Educational aims: to help the participants understand the nature of the water cycle; to show the variety of uses and the importance of water in our lives.
Number of participants needed: 4 - 12 players.
Time required: 1 school double period of roughly 1¼ - 1½ hours.
Contents of package: A4 booklet whose outer sheets (of thin card) are used to prepare the board and other playing materials. The inner sheets

(of paper) constitute an 8-page background reader/instructional booklet for the pupils.

Special facilities required: scissors to prepare playing materials.

Outline of exercise: The board represents the total water cycle which is broken down into 10 different elements. The participants, by answering a series of questions, have to collect cards which allow them to complete the cycle. The game can be extended by getting participants to produce a model water cycle for their own district.

References

1. Bloomer, J (1973) What have simulation and gaming got to do with programmed learning and educational technology? *Programmed Learning & Educational Technology* **10** 4: 224
2. Abt, C C (1968) Games for learning. In Boocock, S S and Schild, E O (eds) *Simulation Games in Learning.* Sage Publications, Beverly Hills
3. Guetzkow, H (1963) *Simulation in International Relations.* Prentice-Hall, Englewood Cliffs, NJ
4. Walker, M (1974) The use of case studies. *Education in Chemistry* **11** 2: 58
5. Percival, F and Ellington, H I (1980) The place of case studies in the simulation/gaming field. In Race, P and Brook, D (eds) *Perspectives on Academic Gaming and Simulation 5.* Kogan Page, London
6. Reid, N (1977) Simulations, games and case studies. *Education in Chemistry* **13** 3: 82
7. Taylor-Byrne, J V (1980) The game of 'Mastermind' as a diagnostic tool for the identification of thinking weaknesses. In Race, P and Brook, D (eds) *Perspectives on Academic Gaming and Simulation 5.* Kogan Page, London
8. Meadows, D H *et al* (1974) *The Limits to Growth.* Pan Books, London
9. Ellington, H I, Addinall, E and Langton, N H (1978) The Bruce Oil Game: a computerized business management game. In Megarry, J (ed) *Perspectives on Academic Gaming and Simulation 1 & 2.* Kogan Page, London.
10. Percival, F (1976) *A Study of Teaching Methods in Chemical Education.* PhD thesis, University of Glasgow
11. Bloom, B S (ed) (1956) *Taxonomy of Educational Objectives. Book 1: Cognitive Domain.* Longman, London
12. Bloom, B S *et al* (1964) *Taxonomy of Educational Objectives. Book 2: Affective Domain.* Longman, London
13. Wentworth, D R and Lewis, D R (1973) A review of research on instructional games and simulations in social science education. *Social Education* May, 432
14. Twelker, P A (1971) Simulation and media. In Tansey, P J (ed) *Educational Aspects of Simulation.* McGraw-Hill, London
15. Gagné, R M (1970) *The Conditions of Learning* (2nd edition). Holt, Rhinehart and Winston, London
16. Ellington, H I and Percival, F (1977) Educating 'through' science using multi-disciplinary simulation games. *Programmed Learning & Educational Technology* **14** 2: 117

17. Garvey, D M (1971) Simulation: a catalogue of judgments, findings and hunches. In Tansey, P J (ed) *Educational Aspects of Simulation.* McGraw-Hill, London

18. Braddock, C (1967) Project 100,000. *Phi Delta Kappan* **48**: 426

19. Daniels, D J (ed) (1975) *New Movements in the Study and Teaching of Chemistry.* Temple Smith, London

20. Thomas, C J (1957) The genesis and practice of operational gaming. *Proceedings of the First International Conference on Operational Research.* Operations Research Society of America, Baltimore

21. Tansey, P J and Unwin, D (1969) *Simulation and Gaming in Education.* Methuen Educational, London

22. Riccardi, F M (1957) *Top Management Decision Simulation: The AMA Approach.* American Management Association, New York

23. Hemphill, J K, Griffiths, D and Frederiksen, N (1962) *Administrative Performance and Personality.* Columbia University Bureau of Publications, New York

24. Kersh, B Y (1962) The classroom simulator. *Journal of Teacher Education* **13** 3: 110

25. Cruickshank, D B (1966) Simulation: new direction in teacher preparation. *Phi Delta Kappan* **48**: 23

26. Walford, R (1968) *Six Classroom Games for Use in Geography Teaching.* Maria Grey College of Education, Twickenham

27. Van der Eyken, W (1968) The game is the thing. *Times Educational Supplement* May 10: 1589

28. Taylor, J L (1971) *Instructional Planning Systems: A Gaming Simulation Approach to Urban Problems.* Cambridge University Press, Cambridge

29. Zoller, U and Timor, Y (1978) Games and simulations for secondary school chemistry teaching: the periodic table. In McAleese, R (ed) *Perspectives on Academic Gaming and Simulation 3.* Kogan Page, London

30. Bloomer, J (1972) *Evaluating an Educational Game.* MEd thesis, University of Glasgow

31. Megarry, J (1977) Circuitron — an electric circuit game. In Megarry, J (ed) *Aspects of Simulation and Gaming.* Kogan Page, London

32. Dowsey, M (1977) Computer simulation of laboratory experiments. In Megarry, J (ed) *Aspects of Simulation and Gaming.* Kogan Page, London

33. Megarry, J (1978) Retrospect and prospect. In McAleese, R (ed) *Perspectives on Academic Gaming and Simulation 3.* Kogan Page, London

34. Hughes, W (1974) A study of the use of computer-simulated experiments in the physics classroom. *Journal of Computer-Based Instruction* **1** 1: 1

35. Lewis, D G (1964) Objectives in the teaching of science. *Educational Research* **7**: 186

36. Maskill, R (1971) The objectives of tertiary chemistry courses. In *Chemical Education at the Tertiary Level.* British Committee on Chemical Education, London

37. Beard, R M (1973) Objectives in tertiary science education. In Billing, D E and Furniss, B S (eds) *Aims, Methods and Assessment in Advanced Science Education.* Heyden, London

38. Billing, D E (1973) Broad aims of chemistry courses. *Chemistry Society Curriculum Subject News Letter* 1: 22

39. Hadden, R A (1975) *Affective Objectives in Chemistry.* MSc thesis, University of Glasgow

40. Brown, S A (1975) *Affective Objectives in an Integrated Science Curriculum.* PhD thesis, University of Stirling

41. Ellington, H I and Percival, F (1977) The place of multi-disciplinary games in school science. *School Science Review* 59 206: 29

42. Percival, F (1977) The development and evaluation of a structured scientific communication exercise. In Hill, P and Gilbert, J (eds) *Aspects of Educational Technology XI.* Kogan Page, London

43. Johnstone, A H, Percival, F and Reid, N (1978) Simulations and games in the teaching of chemistry. In Megarry, J (ed) *Perspectives on Academic Gaming and Simulation 1 & 2.* Kogan Page, London

44. Snow, C P (1959) *The Two Cultures and the Scientific Revolution.* Cambridge University Press, Cambridge

45. Lewis, J L (1978) Science in society. *Physics Education* 13 6: 340

46. Ziman, J (1978) Summary talk given at conference on science, society and education. In Boeker, E and Gibbons, K (eds) *Science, Society and Education.* Free University, Amsterdam

47. Dixon, B (1978) Education for participation. *New Scientist* 79 1117: 530

48. Gosling, W (1978) *Microcircuits, Society and Education.* Council for Educational Technology, London

49. (1978) *Microelectronics: Their Implications for Education and Training.* Council for Educational Technology, London

50. Boeker, E and Gibbons, M (1978) Introduction. *Science, Society and Education.* Free University, Amsterdam

51. Lewis, J L (1978) 'Science and society' education in British schools. In Boeker, E and Gibbons, M (eds) (1978) *Science, Society and Education.* Free University, Amsterdam

52. Ellington, H I, Percival, F, Addinall, E and Smythe, M E (1978) Using simulations and case studies to teach the social relevance of science. In Boeker, E and Gibbons, M (eds) *Science, Society and Education.* Free University, Amsterdam

53. Ellington, H I, Langton, N H and Smythe, M E (1977) The use of simulation games in schools — a case study. In Hills, P and Gilbert, J (eds) *Aspects of Educational Technology XI.* Kogan Page, London

54. Millar, J W L (1979) The Power Station Game: a study. *Physics Education* **14** 1: 34
55. Megarry, J (1975) A review of science games — variations on a theme of rummy. *Simulations and Games* **6** 4: 423
56. (1973) *PHI — Science in Small Schools.* Dept of Education, University of Glasgow
57. McKenzie, J, Elton, L and Lewis, R (eds) (1978) *Interactive Computer Graphics in Science Teaching.* Ellis Horwood, Chichester
58. Hinton, T (1978) Computer assisted learning in physics. *Computers and Education* **2**: 71
59. Tribe, M A and Peacock, D (1973) The use of simulated exercises (games) in biological education at the tertiary level. In Budgett, R and Leedham, J (eds) *Aspects of Educational Technology VII.* Pitman, London
60. Dowdeswell, W and Bailey, C (1978) Simulation in the teaching of a first-year ecology course. In McAleese, R (ed) *Perspectives on Academic Gaming and Simulation 3.* Kogan Page, London
61. Bradley, D (1979) Playing games at school. *IEE News* **42**: 3
62. Ellington, H I and Langton, N H (1979) The Power Station Game Project and its aftermath. *Bulletin of Scottish Curriculum Development Service, Dundee Centre* **15**: 11
63. Reid, N (1977) Games and simulations for O-grade chemistry — a brief report of ongoing research and development work. *SAGSET Journal* **7** 2: 48
64. Reid, N (1979) Simulation approaches in chemistry teaching: update. *Simulation/Games for Learning* **9** 3: 122
65. Reid, N (1978) *Attitude Development through a Science Curriculum.* PhD thesis, University of Glasgow
66. Nasr, M A (1976) *A Study of the Affective Domain in School Science.* PhD thesis, University of Glasgow
67. Addinall, E and Ellington, H I (1978) Hydropower 77. In McAleese, R (ed) *Perspectives on Academic Gaming and Simulation 3.* Kogan Page, London
68. Ellington, H I and Addinall, E (1977) The Multi-Disciplinary Multi-Project Pack — a new concept in simulation gaming. *Programmed Learning & Educational Technology* **14** 3: 213
69. Ellington, H I, Addinall, E and Smith, I H (1977) 'Power for Elaskay' — the Hydroboard's new competition for secondary schools. *Bulletin of Scottish Centre for Mathematics, Science and Technical Education* **12**: 21
70. Ellington, H I and Addinall, E (1978) 'Power for Elaskay' — a learning package on alternative energy resources for use by science teachers. *School Science Review* **59** 209: 747
71. Ellington, H I and Addinall, E (1979) Building case study simulations into the science curriculum as part of structured lessons. *Simulation/Games for Learning* **9** 1: 13
72. Ellington, H I, Addinall, E and Hateley, M C (1980) The 'Project Scotia' Competition. *Physics Education* **15** 4: 221

73. Ellington, H I, Langton, N H and Smythe, M E (1978) The Power Station Game Competition for the Agecroft Trophy. *SAGSET Journal* **8** 2: 56

74. Smythe, M E, Mellor, V A and Ellington, H I (1979) The 1978 Agecroft Trophy Competition. In Megarry, J (ed) *Perspectives on Academic Gaming and Simulation 4*. Kogan Page, London

75. Ellington, H I, Percival, F and Addinall, E (1979) Building science-based educational games into the curriculum. In Race, P and Brook, D (eds) *Aspects of Educational Technology XII*. Kogan Page, London

76. Ellington, H I, Addinall, E, Percival, F and Lewis, J L (1979) Using simulations and case studies in the ASE's 'Science in Society' project. In Megarry, J (ed) *Perspectives on Academic Gaming and Simulation 4*. Kogan Page, London

77. Eglinton, G and Maxwell, J R (1971) Chemsyn — chemical card game 1. *Education in Chemistry* **8** 4: 142

78. Ellington, H I and Addinall, E (1978) North Sea : a board game on the offshore oil industry. In Megarry, J (ed) *Perspectives on Academic Gaming and Simulation 1 & 2*. Kogan Page, London

79. Ellington, H I and Addinall, E (1980) Converting a family board game into an educational package — a case study. In Race, P and Brooks, D (eds) *Perspectives on Academic Gaming and Simulation 5*. Kogan Page, London

80. Ellington, H I, Percival, F and Addinall, E (1979) Simulation-games and case studies — some relationships between objectives and structure. In Page, G T and Whitlock, Q (eds) *Aspects of Educational Technology XIII*. Kogan Page, London

81. Percival, F and Ellington, H I (1978) Fluoridation? — a role-playing simulation game for schools and colleges. *SAGSET Journal* **8** 3: 93

82. Percival, F and Ellington, H I (1979) An attempt to evaluate 'Fluoridation?' — a role-playing simulation exercise. In Megarry, J (ed) *Perspectives on Academic Gaming and Simulation 4*. Kogan Page, London

83. Ellington, H I, Garrow, A G and Muckersie, J R (1978) A simulated public inquiry for use in schools and colleges. In Megarry, J (ed) *Perspectives on Academic Gaming and Simulation 1 & 2*. Kogan Page, London

84. Hooper, R (1977) *National Development Programme in Computer Assisted Learning. Final Report of the Director.* Council for Educational Technology, London

85. Easton, M J, Johnstone, A H and Reid, N (1978) Computer managed chemistry teaching for secondary and tertiary students. *British Journal of Educational Technology* **9** 1: 37

86. Magee, B (1973) *Popper*. Fontana/Collins, Glasgow (Chapter 5)

87. Popper, K R (1959) *The Logic of Scientific Discovery*. Hutchinson, London

88. Popper, K R (1973) *Objective Knowledge: An Evolutionary Approach*. Oxford University Press, Oxford

89. Bloomer, J (1975) Paradigms of evaluation. *SAGSET Journal* **5** 1: 36
90. Shirts, R G (1970) Games students play. *Saturday Review* **53**: 81
91. Clarke, M (1978) *Simulations in the Study of International Relations.* G W and A Hesketh, Ormskirk, Lancashire
92. Tansey, P J (1973) Evaluation, statistics both slammed. *Simulation/Gaming News* **1** 5: 4
93. Walford, R (1975) Evaluation. *SAGSET Journal* **5** 1: 20
94. Percival, F (1978) Evaluation procedures for simulation/gaming exercises. In McAleese, R (ed) *Perspectives on Academic Gaming and Simulation 3.* Kogan Page, London
95. Scarfe, N V (1971) Games, models and reality in the teaching of geography in school. *Geography* **56**: 191
96. Vaughan, K (1977) Evaluation of chemical card games as learning aids. In Hills, P and Gilbert, J *Aspects of Educational Technology XI.* Kogan Page, London
97. Likert, R (1932) A technique for the measurement of attitudes. *Archives of Psychology* **140**: 55
98. Osgood, C E, Suci, G J and Tannenbaum, P H (1957) *The Measurement of Meaning.* University of Illinois Press
99. Vesper, K H and Adams, J L (1969) Evaluating learning from the case method. *Engineering Education* October, 104
    Handy, J and Johnstone, A H (1973) Science education — what is left? In *The Discipline of Chemistry — Its Place in Education.*
100. The Chemical Society (Education Division), London

Bibliography

Those interested in reading further may find the following books and articles of interest.

Beech, G (ed) (1978) *Computer Assisted Learning in Science Education.* Pergamon Press, Oxford

Bloomer, J (1972) *Evaluating an Educational Game.* MEd thesis, University of Glasgow

Bloomer, J (1973) What have simulation and gaming got to do with programmed learning and educational technology? *Programmed Learning & Educational Technology* 10 4:224

Boocock, S S and Schild, E O (eds) (1968) *Simulation Games in Learning.* Sage Publications, Beverly Hills

Clarke, M (1978) *Simulations in the Study of International Relations.* G W and A Hesketh, Ormskirk, Lancashire

Daniels, D J (ed) (1975) *New Movements in the Study and Teaching of Chemistry.* Temple Smith, London

Ellington, H I and Addinall, E (1979) Building case study simulations into the science curriculum as part of structured lessons. *Simulation/Games for Learning* 9 1: 13

Ellington, H I, Addinall, E, Percival, F and Lewis, J L (1979) Using simulations and case studies in the ASE's 'Science in Society' project. In Megarry, J (ed) *Perspectives on Academic Gaming and Simulation 4.* Kogan Page, London

Ellington, H I, Langton, N H and Smythe, M E (1977) The use of simulation games in schools — a case study. In Hills, P and Gilbert, J (eds) *Aspects of Educational Technology XI.* Kogan Page, London

Ellington, H I and Percival, F (1977) Educating 'through' science using multi-disciplinary simulation games. *Programmed Learning & Educational Technology* 14 2: 117

Ellington, H I and Percival, F (1977) The place of multi-disciplinary games in school science. *School Science Review* 59 206: 29

Ellington, H I, Percival, F and Addinall, E (1979) Building science-based educational games into the curriculum. In Race, P and Brook, D (eds) *Aspects of Educational Technology XII.* Kogan Page, London

Ellington, H I, Percival, F and Addinall, E (1979) Simulation-games and case studies — some relationships between objectives and structure.

In Page, G T and Whitlock, Q (eds) *Aspects of Educational Technology XIII*. Kogan Page, London

Ellington, H I, Percival, F, Addinall, E and Smythe, M E (1978) Using simulations and case studies to teach the social relevance of science. In Boeker, E and Gibbons, M (eds) *Science, Society and Education*. Free University, Amsterdam

Hooper, R (1977) *National Development Programme in Computer Assisted Learning. Final Report of the Director*. Council for Educational Technology, London

Johnstone, A H, Percival, F and Reid, N (1978) Simulations and games in the teaching of chemistry. In Megarry, J (ed) *Perspectives on Academic Gaming and Simulation 1 & 2*. Kogan Page, London

Lewis, J L (1978) 'Science and society' education in British schools. In Boeker, E and Gibbons, M (eds) (1978) *Science, Society and Education*. Free University, Amsterdam

McKenzie, J, Elton, L and Lewis, R (eds) (1978) *Interactive Computer Graphics in Science Teaching*. Ellis Horwood, Chichester

Megarry, J (1975) A review of science games — variations on a theme of rummy. *Simulations and Games* 6 4: 423

Megarry, J (1978) Retrospect and prospect. In McAleese, R (ed) *Perspectives on Academic Gaming and Simulation 3*. Kogan Page, London

Nasr, M A (1976) *A Study of the Affective Domain in School Science*. PhD thesis, University of Glasgow

Percival, F (1976) *A Study of Teaching Methods in Chemical Education*. PhD thesis, University of Glasgow

Percival, F (1978) Evaluation procedures for simulation/gaming exercises. In McAleese, R (ed) *Perspectives on Academic Gaming and Simulation 3*. Kogan Page, London

Percival, F and Ellington, H I (1980) The place of case studies in the simulation/gaming field. In Race, P and Brook, D (eds) *Perspectives on Academic Gaming and Simulation 5*. Kogan Page, London

Reid, N (1977) Games and simulations for O-grade chemistry — a brief report of ongoing research and development work. *SAGSET Journal* 7 2: 48

Reid, N (1977) Simulations, games and case studies. *Education in Chemistry* 13 3: 82

Reid, N (1978) *Attitude Development through a Science Curriculum*. PhD thesis, University of Glasgow

Roebuck, M (1978) Simulation games and the teacher as an adaptive interventionist. In McAleese, R (ed) *Perspectives on Academic Gaming and Simulation 3*. Kogan Page, London

Spencer, J (1977) Games and simulations for science teaching. *School Science Review* 58 204: 397

Tansey, P J (ed) (1971) *Educational Aspects of Simulation.* McGraw-Hill, London

Tansey, P J and Unwin, D (1969) *Simulation and Gaming in Education.* Methuen Educational, London

Taylor, J L and Walford, R (1978) *Learning and the Simulation Game.* Open University Press, Milton Keynes

Teather, D C B (1978) Simulation and games. In Unwin, D and McAleese, R (eds) *Encylopaedia of Educational Media Communications and Technology.* Macmillan, London

Walford, R (1975) Evaluation. *SAGSET Journal* **5** 1: 20

Walker, M (1974) The use of case studies. *Education in Chemistry* **11** 2: 58

Wentworth, D R and Lewis, D R (1973) A review of research on instructional games and simulations in social science education. *Social Education* May, 432

Index to Part 1

Index to Part 2